PHOENIX WILDFLOWERS

PHOENIX
Wildflowers

A Field Guide to the Common Shrubs, Herbs, Cacti and Ferns of the Phoenix Region

ARIZONA HEDGEHOG CACTUS
(*Echinocereus arizonicus*)
Illustration by Margaret Neilson Armstrong

FRANCES L. HAMILTON
STEVE CHADDE

Phoenix Wildflowers
A Field Guide to the Common Shrubs, Herbs, Cacti and Ferns of the Phoenix Region

Frances L. Hamilton and Steve Chadde

ISBN 978-1951682910
Printed in the United States of America

A Pathfinder Field Guide
Published by Orchard Innovations, Mountain View, Arkansas
Author email: *steve@orchardinnovations.com*

VERSION 1.0 01/27/2025

CONTENTS

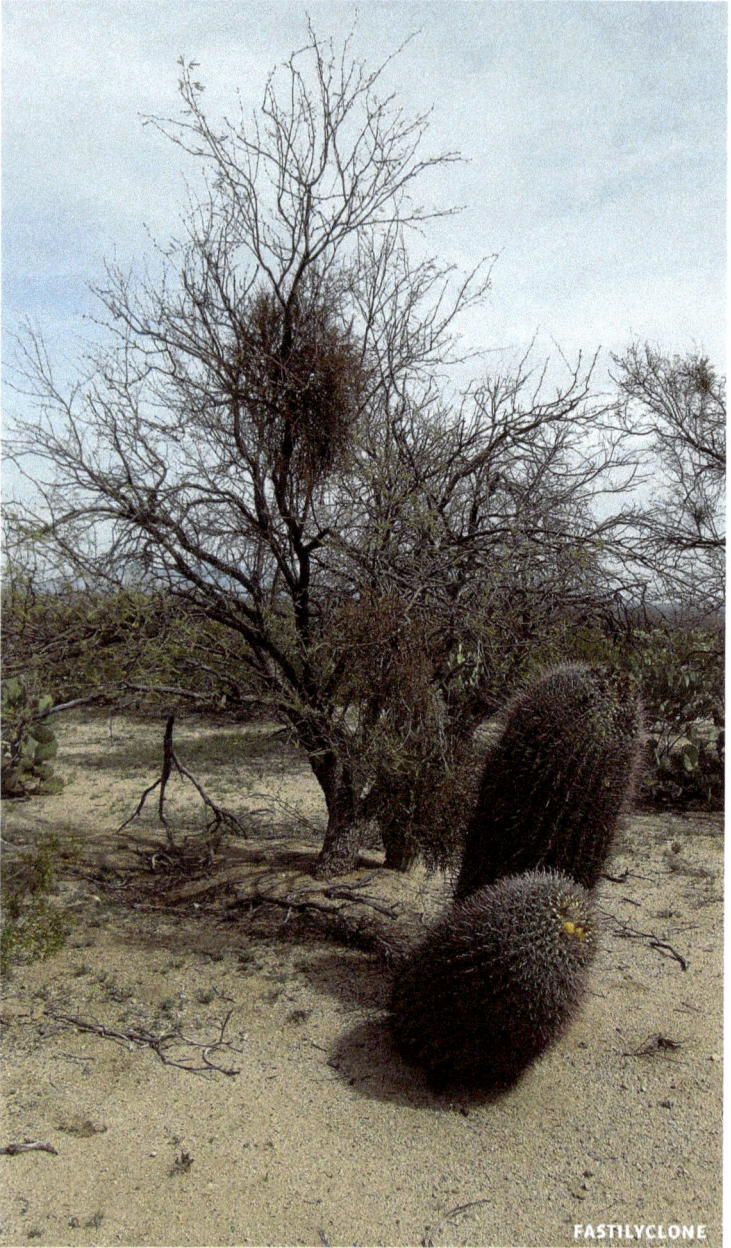

A Mesquite tree (*Prosopis*) with Desert Mistletoe (*Phoradendron*), two large plants of Fishhook Barrel Cactus (*Ferocactus*), with Prickly-Pear Cactus (*Opuntia*)in the background; Sahuarita, Arizona.

INTRODUCTION

"THE DESERT A GARDEN?" The aim of this book is to introduce residents and visitors to the Phoenix, Arizona, region the large and diverse flora present in this part of the Sonoran Desert. No month of the year finds the desert devoid of flowers and leafy shrubs though some months are favored: early fall and spring find the desert carpeted with a tapestry of interwoven yellow, orange, red, lavender, blue, and pink blossoms. Some of the flowers are large, brilliant, and flamboyant, but the desert is unique with its numerous plants with only very tiny flowers.

Phoenix, the hub from which each road leads to a different and equally enchanting spot, is bounded on the north, south, and east by mountains. On the north lies Camelback, a characteristic and well-known mountain peak. Slightly to the west of Camelback lies Piestewa Peak, the highest and best known of a small group of mountains. To the east and looking strangely like a giant's rock garden, so carelessly are the rocks piled, is Papago Park. South of Phoenix, across the Salt River, are the South or Salt River Mountains.

This is the location of one of the largest city parks in the United States. The area of the park is approximately 16,000 acres. In these are present many things of both historical and biological interest including plant and animal life and fascinating hieroglyphics carved by Indians ages ago.

Those hills which from a distance seem so barren are the home of a surprising number of plant species. The slopes are rocky, there is not a great amount of soil in which plants can grow, yet they seem to thrive. The predominating plants in the region are shrubs, trees and cacti. Creosotebushes (*Larrea tridentata*, p. 71) with their glossy green leaves and bright yellow flowers are abundant. Closely associated with Creosotebush are two small shrubs, Rabbitbush (*Ambrosia deltoidea*, p. 153) and Burroweed (*Ambrosia dumosa*, p. 154). The Rabbitbush has small, gray, wrinkled leaves, the Burroweed has leaves that are also small and gray but deeply lobed. Brittlebush (*Encelia farinosa*, p. 32) is another of the very common shrubs. In the spring it is covered with many yellow flowers borne on slender stems above the silvery leaved plants.

The most conspicuous plants are the Ocotillos (*Fouquieria splendens*, p. 143) and the cacti. The Ocotillo with its long, slender, whip-like branches and clusters of bright red or orange-red flowers is particularly attractive in the early spring. The cacti are always a fascinating study. The Saguaro or Giant Cactus (*Carnegiea gigantea*, p. 89), the state flower of Arizona, is limited to a small area in southern Arizona, northern Sonora, Mexico, and several locations in extreme southeastern California. The flowers are borne in crowns at the tips of the branches. The blooms always open on the southwest side of the plants first. The fruits are edible and were used by the Indians in making jams. These giants are many years old, and young plant plants grow slowly—at six or seven years old they are only an inch high.

The most common cacti are the Chollas. These bear small, nearly inconspicuous spines called glochids. Glochids are armed with fine barbs directed backwards, much like fish hooks. It is almost impossible to pull them out when once they get into the flesh.

Trees are quite abundant especially down the washes where there is more water than the rocky hills afford. The Palo-Verde (*Parkinsonia*, p. 55) is noticeable for its great masses of yellow flowers and the soft, green branches from which it gets its name. The Mesquite (*Prosopis*, pp. 94, 95) was useful to the Indians. From the beans they made flour for mush, bread, and tortillas. They also extracted sugar from the beans. Other trees are the Desert Ironwood (*Olneya tesota*, p. 119), Catclaw Acacia (*Senegalia greggii*, p. 57), and Elephant-Tree (*Bursera microphylla*, p. 87). The Elephant-Tree is unique, being found only in a very small area in southern Arizona, southern California, and Mexico.

The smaller plants, though not so conspicuous, are much more numerous. The Desert Plantain or Indian Wheat (*Plantago ovata*, p. 100) is the most common. In appearance it closely resembles a small grass and is usually so abundant and the plants so close together that the desert has its own lawn. White Easterbonnets (*Eriophyllum lanosum*, p. 78), a tiny plant barely two inches high, is conspicuous with its yellow and white flower heads. The slender Goldfields (*Lasthenia gracilis*, p. 36), the tiny Desert Threadplant (*Nemacladus rubescens*, p. 91), the Wild Carrot (*Daucus pusillus*, p. 74), Lupines (*Lupinus*, pp. 117, 118), Gilias (*Gilia*, p. 135), Morning Brides (*Chaenactis stevioides*, p. 77), Combseeds (*Pectocarya*, p. 82), Chias (*Condea, Salvia*, pp. 125, 126), and Fleabanes (*Erigeron*, p. 106), all have their place in the desert flora.

The outstanding color of the desert flowers is yellow, with white and orange standing next in importance, while pinks, blues, lavenders and reds, though present, are not abundant. Yellow flowers seem to thrive best in the sun and are the most numerous in the desert.

A few of our plants might be called "flowerless"—these are the ferns, spikemosses and lichens. There are only two small ferns of note in the area: the Copper Fern (*Bommeria hispida*, p. 165) and the Lip Fern (*Myriopteris pringlei*, p. 166). Neither is very abundant and can be found only upon close search at the bases of large rocks. The spikemoss, *Selaginella*, (p. 168) forms the basis of the beautiful rock gardens. It is very abundant and though having a dead appearance in dry weather, becomes dark green in rainy weather. Some of the rocks have the appearance of having been painted green, yellow, orange, or red. This "paint" is formed by crustose lichens. Lichens are plants made up of algae and fungi living in close relationship. In plant succession from barren rocks to a plant growth of any kind, lichens are the first plants to appear. They grow as epiphytes on dry rocks, gradually causing them to disintegrate until there is enough soil formed for the larger types of plants to grow.

Desert plants as a whole are adapted to protect themselves from various injurious elements; their two most important enemies being heat and loss of moisture. There is not an abundance of rainfall, and there are no large bodies of water or fogs from which the plants can obtain the necessary amount of moisture. The cacti have the best protection. They are covered by an armor of spines which keeps off the sun, lessening the amount of water lost through transpiration. They have are no leaves except on very young plants and joints, and these soon fall off. The stem of the cactus is green and contains the chlorophyll which enables it to manufacture food. The center is fibrous and filled with large cells capable of holding a large supply of water, and the root system is shallow, the roots being near the surface of the ground in order to take advantage of any rains. All desert plants have a reduced leaf area; for example the Palo-Verde has very small leaves, the green trunk helping in the manufacture of the food. Many desert plants also have thickened leaves and breathing pores which are deeply set and can be covered by surrounding cells during the day. The small annual plants grow during the spring and fall months and thus have no need of special protection.

The odors and tastes of many plants are also of protective value. For example, Odora (*Porophyllum gracile*, p. 107) has an extremely disagreeable odor when crushed, and Rockflower (*Crossosoma bigelovii*, p. 92) has a bitter taste if chewed. Cactus blossoms as a rule have no decided odor, but some, like the Night-Blooming Cereus (*Peniocereus*), have a heavy odor.

On warm spring days many insects are attracted to the plants. Bees, butterflies, wasps, beetles, grasshoppers, and flies of various types are seen. The Mesquite (*Prosopis*, pp. 94. 95) is an excellent honey plant and when in bloom is always visited by many bees. The Flatcrown Buckwheat (*Eriogonum deflexum*, p. 136) is also a good honey plant.

USING THIS BOOK

This Flora has been written in semi-scientific style so that the layperson as well as the botanist may know and learn to appreciate our desert flowers. Plants are first grouped by their flower color. Within each color group, plants are arranged alphabetically by the scientific name of their family (e.g., Asteraceae, Aster Family), then within the family, by their genus name (e.g., *Brickellia*, Brickellbush). A key to each plant family treated in the book is provided (see p. 11), followed by brief family descriptions and additional keys to the genera and species. An illustrated Glossary (p. 169) defines botanical terms used in the text, and is followed by an index to the common and scientific names of the more than 150 plants described.

ACKNOWLEDGMENTS

I would like to give my sincere thanks to the many people who have made their photographs available under the appropriate Creative Commons license allowing for commercial use. Nearly all photographs were obtained from the *www.inaturalist.org* website, the leading community-driven site for sharing ecological data. Photographer names (as provided on inaturalist) are listed on each image. This book would not have been possible without their efforts to document our diverse flora.

Acknowledgment is also given to the Biota of North America Program (BONAP) for use of their data to confirm species presence within the Phoenix region (see *www.bonap.org*). Current taxonomic names largely follow those of BONAP and of the 'Plants of the World Online' database maintained by the Royal Botanic Gardens, Kew (see *powo.science.kew.org*).

—STEVE CHADDE

KEY TO THE FAMILIES

Use the family key below to place an unknown plant in its botanical family. This key is followed by a brief description of each family's characteristics, and additional keys to the genera and species included in this book. Note that not all plant species of the Phoenix area are described—there are over 1,600 different plant species reported from Maricopa County alone.

1   Plants bearing no flowers 2
1   Plants bearing flowers 3
2   Ferns . . . . PTERIDACEAE
2   Low moss-like plants . . . . SELAGINELLACEAE
3   Flowers dioecious, petals numerous, scale-like; flowers resembling small cones. . . . . EPHEDRACEAE
3   Flowers perfect, monoecious or dioecious, petals few, not scalelike; flowers not cone-like 4
4   Flower parts in three's; leaves slender, parallel-veined 5
4   Flower parts in four's or five's (rarely three's, Burseraceae); leaves various, net-veined. 6
5   Corolla composed of small green glumes; leaves 2 to 6 inches long. . . . . POACEAE
5   Corolla colored, showy; leaves 12 to 15 inches long, somewhat twisting. . . . . ASPARAGACEAE
6   Corolla absent or inconspicuous; calyx present, sometimes showy, or absent. 7
6   Corolla and calyx present 13
7   Calyx absent; corolla showy . . . . OLEACEAE
7   Calyx present, sometimes showy. 8
8   Ovary superior 9
8   Ovary inferior or appearing so by the closely fitting calyx 12
9   Pistils several to many, distinct; calyx showy . . . . . RANUNCULACEAE
9   Pistils one, one to several-celled. 10
10 Plants with milky juice in the stems . . . . EUPHORBIACEAE
10 Plants without juice in the stems 11
11 Stipules, if present, sheathing the stem; sepals three to six . . . . POLYGONACEAE
11 Stipules absent; sepals mostly five . . . . AMARANTHACEAE
12 Herbs; flowers showy; leaves present . . . . NYCTAGINACEAE
12 Parasites; flowers inconspicuous; leaves scalelike. . . . . SANTALACEAE
13 Corolla of separate petals 14
13 Corolla with petals more or less united 28
14 Ovary superior 15
14 Ovary inferior or more or less so 25

15 Trees; leaves with a strong pepper-tree odor when crushed; petals three, tiny . . . . **BURSERACEAE**

15 Herbs or shrubs; petals five, usually showy **16**

16 Stamens more than ten in number **17**

16 Stamens ten or less in number **19**

17 Pistils separate and distinct; filaments of stamens not united **18**

17 Pistils united at base into a lobed or beaked ovary; filaments of stamens united . . . . **MALVACEAE**

18 Shrubs; flowers cream-color, rose-like . . . . **CROSSOSOMATACEAE**

18 Small annuals; flowers bright orange . . . . **PAPAVERACEAE**

19 Ovaries five, separate or partly united **20**

19 Ovary one, the styles and stigmas one to several **21**

20 Low spreading herbs; carpels with tails 1½ to 2 inches long; flowers purple of pink-purple . . . . **GERANIACEAE**

20 Shrubs or herbs; carpels rounded and hairy or with short stout spines; flowers yellow or orange . . . . **ZYGOPHYLLACEAE**

21 Ovary with one cell and one placenta **22**

21 Ovary with one or more cells and styles, and two or more placentae and stigmas **23**

22 Pods beans; petals large, regular or irregular; stamens ten **FABACEAE**

22 Pods ovoid, covered with dark red prickles; petals reduced to small appendages; stamens three to four . . . . **KRAMERIACEAE**

23 Perennial vine; fruit a samara . . . . **MALPIGHIACEAE**

23 Annual herbs; fruit a pod **24**

24 Ovules attached at the center or bottom of the ovary; stamens equal; stems with dark sticky bands on the internodes . **CARYOPHYLLACEAE**

24 Ovules attached on two placentae; the six stamens unequal; stems without sticky bands . . . . **BRASSICACEAE**

25 Stamens more than ten in number **26**

25 Stamens less than ten in number **27**

26 Plants spiny; leaves absent or deciduous soon . . . . **CACTACEAE**

26 Plants not spiny; leaves persisting for the season; stamens five in some species . . . . **LOASACEAE**

27 Seeds more than one in each cell; stamens four to eight; fruit slender 1 to 2 inches long; petals four . . . . **ONAGRACEAE**

27 Seeds only one in each cell; stamens five; fruit not exceeding ⅛ inch in length; petals five . . . . **APIACEAE**

28 Ovary superior **29**

28 Ovary inferior **37**

29 Corolla regular **30**

29 Corolla irregular **35**

30 Tall spiny shrubs, many heavy stems from the base; flowers scarlet . . . . **FOUQUIERIACEAE**

30 Low annuals, branched more or less throughout; the stems not spiny **31**

31 Ovary one or two-celled **32**

31 Ovary three or four-celled **34**
32 Flowers in a dense spike; petals four; leaves all basal . . . .
. . . . **PLANTAGINACEAE**
32 Flowers not in a dense spike; petals five; leaves not all basal **33**
33 Styles two or occasionally only one . . . . former **HYDROPHYLLACEAE**,
. . . . now placed as a subfamily of the **Boraginaceae**
33 Styles one or none. . . . . **SOLANACEAE**
34 Ovary three parted; style one . . . . **POLEMONIACEAE**
34 Ovary four-celled; styles two . . . . **BORAGINACEAE**
35 Ovary two-celled **36**
35 Ovary four-lobed; plants mostly aromatic. . . . . **LAMIACEAE**
36 Shrub; flowers tubular, very slender, orange-red . . . . . **ACANTHACEAE**
36 Annual herbs; flowers not slender, yellow or purplish . . . .
. . . . . former **SCROPHULARIACEAE** (**Orobanchaceae, Phrymaceae**)
37 Stamens free, not united; plants rough scabrous . . . . **RUBIACEAE**
37 Stamens united by anthers or filaments; plants not scabrous. **38**
38 Flowers solitary; the filaments of the stamens united at the apices . .
. . . . **CAMPANULACEAE**
38 Flowers numerous, in heads; stamens united by their anthers; fruit a
one-seeded achene. . . . . **ASTERACEAE**

ACANTHACEAE Acanthus Family
Shrub; leaves opposite; flowers perfect, irregular, in few flowered racemes; calyx of five somewhat united sepals; corolla of five united petals, two-lipped; stamens slightly exerted; fruit a capsule.
Hummingbird-Bush (*Justicia californica*, p. 142)

AMARANTHACEAE Amaranth Family
Shrub or annual herbs (ours); leaves opposite or whorled; flowers small, in pairs or close spikes; ovary one-celled; fruit a one-seeded utricle. Includes members of the former Goosefoot Family (Chenopodiaceae), that family now incorporated within the Amaranthaceae.

1 Prostrate annual herb. . . . . **Woolly Honeysweet**
. . . . (*Tidestromia lanuginosa*, p. 29)
1 Plants larger, shrubs or herbs. **2**
2 Shrub; leaves obovate, dull gray-green and scurfy. . . . .
. . . . **Desert Orach** (*Atriplex polycarpa*, p. 149)
2 Annual herb; leaves awl-shaped, rigid; plant becoming a tumbleweed
. . . . **Russian-Thistle** (*Salsola tragus*, p. 150)

APIACEAE Carrot Family
Ours annual herbs; leaves various; flowers small, in umbels; calyx of five teeth or wanting; petals and stamens five; ovary inferior; carpels two.

1   Stems trailing; leaves broad, shallowly seven-lobed; sparsely stellate hairy; flowers few in the axils of the leaves. . . . . **Hoary Bowlesia**
. . . . (*Bowlesia incana*, p. 73)
1   Stems erect; leaves finely dissected; hirsute or glabrous; flowers numerous, in umbels at the spices of the stems. 2
2   Leaves compound pinnate, the divisions very tiny; short hirsute; prickles on the fruits finely barbed . . . . **American Wild Carrot**
. . . . (*Daucus pusillus*, p. 74)
2   Leaves compound pinnate, the divisions linear; glabrous; prickles on the fruits hooked . . . . **Bristly-Fruit Scaleseed**
. . . . (*Spermolepis echinata*, p. 75)

ASPARAGACEAE Asparagus Family
Plants growing from bulbs. Leaves long, slender, partly clasping at the base. Flowers in clusters at the apex of the long slender stem. Sepals three, petal-like; petals three; stamens three; pistil superior, three-parted.
   Bluedicks (*Dipterostemon capitatus*, p. 104)

ASTERACEAE Aster Family
Annual and perennial herbs, shrubs; leaves various; flowers small, in dense heads. Flowers of two types, ray flowers and disk flowers. The heads may be composed of both ray and disk flowers or of all ray flowers or all disk flowers, or the flowers may be inconspicuous and green. Petals five, united into a tube; sepals absent or modified into the pappus; stamens five, the anthers united; pistil inferior, two-celled, only one developing into an achene.

1   Flower heads inconspicuous and greenish or yellow-green, dioecious or monoecious; fruits small "burs." 2
1   Flower heads composed all of ray flowers, all of disk flowers or of both ray and disk flowers 6
2   Plants with the leaves finely dissected. 3
2   Plants with the leaves entire or at most with the margins toothed or scalloped. 4
3   Annuals or biennials, one or two stems from the base; leaves 2 to 3 inches long . . . . **Slimleaf Bursage** (*Ambrosia confertiflora*, p. 152)
3   Shrubs, branched throughout; leaves ½ to ¾ inches long . . . .
. . . . **Burroweed** (*Ambrosia dumosa*, p. 154)
4   Plants thickly branched throughout, soft looking; leaves slender, linear, entire . . . . **Cheesebush** (*Ambrosia salsola*, p. 155)
4   Plants coarse; leaves large, coarse. 5
5   Plants 2 to 4 feet high, growing in stream beds; leaves 4 to 6 inches long . . . . . . **Big Bursage** (*Ambrosia ambrosioides*, p. 151)
5   Plants 1 to 2 feet high, growing abundantly on the slopes; leaves 1 to 1½ inches long, gray-green. . . . . **Rabbitbush**
. . . . (*Ambrosia deltoidea*, p. 153)

6 Flower heads composed of both ray and disk flowers (Aster type) . 7
6 Flower heads composed of all ray flowers or all disk flowers 18
7 Flowers yellow, orange-yellow or brown 8
7 Flowers white or lavender 15
8 Shrubs; leaves pubescent 9
8 Annuals or biennials; leaves pubescent but only finely so, or glabrous
   10
9 Plants forming a rounded clump; leaves so densely pubescent as to
   appear silvery gray. . . . . **Brittlebush** (*Encelia farinosa*, p. 32)
9 Plants forming a loose irregular clump; leaves dark green, finely pubescent. . . . . **Parish's Goldeneye** (*Bahiopsis parishii*, p. 30)
10 Flower heads 2 to 2½ inches across, rays orange, disk brown; leaves
    and stems rough hairy. . . . . **Common Sunflower**
    . . . . (*Helianthus annuus*, p. 33)
10 Flower heads 1½ inches across or less, both ray and disk flowers the
    same color. 11
11 Plants very slender, usually only one stem from the base; leaves linear;
    flower heads ½ inch across . . . . **Goldfields** (*Lasthenia gracilis*, p. 36)
11 Plants coarse, branched; leaves not linear; flower heads 1 to 1½ inches
    across 12
12 Leaves pinnatifid, densely white wooly, in basal clusters; flower heads
    with several rows of ray flowers . . . . **Desert Marigold**
    . . . . (*Baileya multiradiata*, p. 31)
12 Leaves not deeply divided, scarcely pubescent, not in basal clusters;
    flower heads with single rows of ray flowers 13
13 Leaves petiolate, gray pubescent . . . . **Cowpen Daisy**
    . . . . (*Verbesina encelioides*, p. 43)
13 Leaves sessile, clasping or perfoliate, green 14
14 Stems and leaves glabrous; leaves perfoliate. . . . **Lemmon's Ragwort**
    . . . . (*Senecio lemmoni*, p. 39)
14 Stems and leaves glandular hairy and with a soft dense pubescence;
    leaves clasping . . . . . **Camphorweed** (*Heterotheca subaxillaris*, p. 34)
15 Flower heads white with yellow disks. 16
15 Flowers lavender or pinkish. 17
16 Tiny white wooly annuals 2 to 3 inches high . . . .
    . . . . **White Easterbonnets** (*Eriophyllum lanosum*, p. 78)
16 Perennials, 12 to 15 inches high, dark green . . . .
    . . . . **Crowfoot Rock-Daisy** (*Galinsogeopsis coronopifolia*, p. 79)
17 Flower heads ¾ inch across, rays about 50, ¼ inch long, slender, pur-
    ple or lavender; disk yellow . . . . **Spreading Fleabane**
    . . . . (*Erigeron divergens*, p. 106)
17 Flower heads ¼ inch across, rays about 20, ⅛ inch or less in length,
    pinkish, disk pink . . . . **Desert American-Aster**
    . . . . (*Symphyotrichum expansum*, p. 109)
18 Flowers composed only of disk flowers (rayless type) 19
18 Flower heads composed only of ray flowers (Chicory type) 26

19 Flowers purplish 20
19 Flowers white, cream colored or yellow 21
20 Leaves linear; involucral bracts six to seven, about ⅛ inch wide, ½ inch long, purple and with elongated purple glands . . . .
. . . . **Odora** (*Porophyllum gracile*, p. 107)
20 Leaves small, ovate-acuminate; involucral bracts many, barely 1/16 inch wide, ¼ inch long, green with thin purple lines . . . .
. . . . **Coulter's Brickellbush** (*Brickellia coulteri*, p. 105)
21 Flowers yellow 22
21 Flowers white or cream-colored 24
22 Plants annuals, 2 to 4 inches tall, leaves densely white woolly . . . .
. . . . **Yellowdome** (*Trichoptilium incisum*, p. 41)
22 Shrubs, 18 to 36 inches tall, leaves not white woolly 23
23 Plants resinous and coarse, leaves linear; flower heads small, very numerous, flowers regular . . . . **Southern Goldenbush**
. . . . (*Isocoma pluriflora*, p. 35)
23 Leaves broad, glossy green; flower heads large, ½ inch across, numerous but not crowded; flowers bilabiate . . . . **American Threefold**
. . . . (*Trixis californica*, p. 42)
24 Plants shrubs 3 to 7 feet high; many branches from the base; leaves small, linear; flowers very numerous, in large clusters, cream-white .
. . . . **Desert Broom** (*Baccharis sarothroides*, p. 76)
24 Plants annuals, 3 to 15 inches high; flowers few 25
25 Plants 8 to 15 inches high, sparsely pubescent; leaves with very slender divisions; flower heads ½ inch across . . . . **Desert Pincushion**
. . . . (*Chaenactis stevioides*, p. 77)
25 Plants 3 to 5 inches high; densely white cottony; flower heads resembling small bolls of cotton . . . . **Woolly-Head Neststraw**
. . . . (*Stylocline micropoides*, p. 81)
26 Flowers pink, rays few, in a single row . . . . **Wire-Lettuce**
. . . . (*Stephanomeria pauciflora*, p. 108)
26 Flowers white or yellow, rays numerous, in several rows 27
27 Flowers all white; involucral bracts few, these curling back from the flower head 28
27 Flowers all yellow . . . . **Desert Chicory**
. . . . (*Rafinesquia neomexicana*, p. 80)
28 Plants common in cultivated places; leaves 3 to 5 inches long, irregularly pinnately lobed, at the base broadened and clasping the stems .
. . . . **Prickly Sow-Thistle** (*Sonchus asper*, p. 40)
28 Plants not found in cultivated places; leaves 1 to 3 inches long 29
29 Flower heads about ½ inch across, not attractive; in seed the "puff-ball" very showy, 1½ inches across; involucral bracts 1 inch long . . . .
. . . . **Silver Puffs** (*Microseris lindleyi*, p. 38)
29 Flower heads 1 inch across, showy, involucral bracts ¼ inch long . . .
. . . . **Desert Dandelion** (*Malacothrix fendleri*, p. 37)

BORAGINACEAE Borage Family

Rough hairy annuals or glabrous, glaucous perennials; leaves simple, alternate; flowers in cymes, these often helicoid or the flowers in clusters; flowers regular, calyx of five united persistent sepals; corolla tubular, five-lobed; stamens five, adnate to the corolla tube; style one; ovary deeply four-lobed; fruit of two to four seed-like nutlets.

1   Plants low growing and somewhat spreading; leaves thick and gray-green, stems smooth and soft to the touch; flowers pale lavender to whitish . . . . **Alkali Heliotrope** . . . . (*Heliotropium curassavicum*, now placed in own family, ... Heliotropiaceae, see p. 123)

1   Plants growing erect or nearly so; leaves and stems coarsely hairy; flowers yellow or cream-white **2**

2   Plants 12 to 24 inches high, coarse; long hispid; flowers orange-yellow . . . . **Rancher's Fiddleneck** (*Amsinckia intermedia*, p. 44)

2   Plants 3 to 12 inches tall, not coarse; short hispid. **3**

3   Plants with the stems spreading; leaves ½ to ¾ inches long, very slender; pods splitting and the four toothed parts curling backwards . . . . **Broad-Fruit Combseed** (*Pectocarya platycarpa*, p. 82)

3   Plants erect; leaves 2 to 3 inches long, ¼ inch wide; stems, leaves and roots with a dark red juice. . . . . **Arizona Popcorn-Flower** . . . . (*Plagiobothrys arizonicus*, p. 83)

BRASSICACEAE Mustard Family

Annual herbs; leaves alternate; flowers in terminal racemes; sepals four; petals four; stamens six, two short and four long; fruit a pod.

1   Flowers white **2**

1   Flowers yellow. **4**

2   Plants coarse, 18 to 24 inches tall; flowers ⅜ inch across; pods ½ inch across. . . . . **Touristplant** (*Dimorphocarpa wislizeni*, p. 84)

2   Plants slender, 6 to 12 inches tall; flowers to ⅛ inch across; pods not exceeding ¼ inches across **3**

3   Leaves long and slender, margins entire; pods not dehiscent, flat, broadly elliptical, the wing margins perforated . . . . **Mountain Fringepod** (*Thysanocarpus laciniatus*, p. 86)

3   Leaves short, oblanceolate, margins shallowly toothed; pods dehiscent, flattened, not perforated . . . . **Hairy-Pod Pepperwort** . . . . (*Lepidium lasiocarpum*, p. 85)

4   Leaves entire; flowers ⅜ inch across; pods globose, 3/16 inch long . . . . **Bladder Pod** (*Physaria gordoni*, p. 45)

4   Leaves deeply lobed; flowers ⅛ to 3/16 inch across; pods linear, 1 to 2 inches long . . . . **London Rocket** (*Sisymbrium irio*, p. 46)

BURSERACEAE Frankincense Family

Tree mostly 4 to 8 feet tall; leaves pinnately divided to the midrib; sepals three; petals three; stamens six; ovary three-celled.

Elephant -Tree (*Bursera microphylla*, p. 87)

CACTACEAE Cactus Family

Succulent, low growing to tree-like; no leaves except on young stems, then they are early deciduous. Stems with clusters of spines; flowers showy; sepals many, grading into the numerous petals; stamens very many; pistil inferior; style with a number of branches, these with the stigmatic surface; fruit fleshy or dry; seeds numerous.

1   Stems jointed, areoles with glochids (cholla, prickly-pear) 2
1   Stems not jointed (saguaro, pin cushion, barrel, etc.) 6
2   Joints large, flat, disk-shape . . . . **Brown-Spined Prickly-Pear**
. . . . (*Opuntia phaeacantha*, p. 52)
2   Joints slender, round 3
3   Joints entirely covered by spines bearing cream-colored sheaths . . . .
. . . . **Teddy-Bear Cholla** (*Cylindropuntia bigelovii*, p. 49)
3   Joints not entirely covered by spines. 4
4   Spines always more than one; joints stouter, 1¼ inch thick. . . . .
. . . . **Buckhorn Cholla** (*Cylindropuntia acanthocarpa*, p. 47)
4   Spines, at least some of them, solitary, sometimes several, acicular; joints slender, rarely more than ⅜ inch thick 5
5   Joints short, usually at right angles to the branches, 3/16 inch thick.
. . . . **Christmas Cactus** (*Cylindropuntia leptocaulis*, p. 50)
5   Joints longer, ⅜ to ⅝ inch thick, usually at an acute angle to the branches . . **Arizona Pencil Cactus** (*Cylindropuntia arbuscula*, p. 48)
6   Plants with stems ribbed. . . . . **Graham's Nipple Cactus**
. . . . (*Cochemiea grahamii*, p. 113)
6   Plants with stems tuberculate 7
7   Plants 10 to 35 feet high; flowers waxy white . . . .
. . . . **Saguaro** (*Carnegiea gigantea*, p. 89)
7   Plants 6 feet high or less; flowers yellow or purple 8
8   Plants solitary, barrel-like; flowers yellow, spines stout, the central one very strongly hooked . . . . **Bisnaga** (*Ferocactus wislizeni*, p. 51)
8   Plants cespitose, stems cylindric; flowers purple; spines slender, the central one long, not hooked . . . . **Hedgehog Cactus**
. . . . (*Echinocereus engelmannii*, p. 114)

CAMPANULACEAE Bellflower Family

Ours an annual herb; leaves alternate, entire; inflorescence a terminal raceme; hypantheum campanulate, ribbed, adnate to the ovary. Sepals five; corolla conspicuously bilabiate, upper lip two-lobed, lower lip three-lobed; stamens five; the filaments united above; ovary inferior; fruit a many seeded capsule. Now includes former members of **Lobelia Family** (Lobeliaceae).

Desert Threadplant (*Nemacladus rubescens*, p. 91)

CROSSOSOMATACEAE Rockflower Family
Shrubs, leaves alternate; flowers regular; sepals five; petals five; stamens numerous; pistils two or more. The flowers very nearly resemble those of the strawberry.
Ragged Rockflower (*Crossosoma bigelovii*, p. 92)

CARYOPHYLLACEAE Pink Family
Ours an annual herb; inflorescence cymose; calyx five-toothed; petals five; stamens ten; styles three; capsule one-celled; seeds many.
Sleepy Catchfly (*Silene antirhina*, p. 115)

EPHEDRACEAE Mormon-Tea Family
Shrubs with branches which are slender and whip-like and leaves reduced to papery scales. The plants are of two kinds, staminate and pistillate. The flowers on the pistillate plants are small green "cones" about 3/8 inch long, composed of overlapping scales and the flowers on the staminate plants are about ¼ inch long, light yellow and also composed of overlapping scales. The numerous conspicuous yellow stamens are borne in the axils of these scales.
Mormon-Tea (Ephedra viridis, p. 156)

EUPHORBIACEAE Spurge Family
Monoecious or dioecious annual herbs or shrubs with acrid or milky juice in the stems; leaves simple, alternate or opposite; corolla often wanting, involucre resembling a calyx; stamens few; ovary three-celled; fruit a three-celled capsule.

1   Plants annuals, procumbent or nearly so; leaves opposite . . . .
. . . . **Sonoran Sandmat** (*Euphorbia micromeria*, p. 158)
1   Plants shrubby, erect; leaves alternate. **2**
2   Leaves very narrow, linear, not pubescent; flowers inconspicuous. . .
. . . . **Beetle Spurge** (*Euphorbia eriantha*, p. 157)
2   Leaves linear-lanceolate, covered with an appressed silky pubescence; flowers ¼ inch across, pale yellow . . . . **Yuma Silverbush**
. . . . (*Argythamnia serrata*, p. 93)

FABACEAE Pea Family
Herbs, shrubs, trees; leaves divided, palmate or pinnate; flowers regular or irregular; sepals five, united at the base; petals five, one banner, two wings and two keels in the irregular flowers; stamens ten, usually nine united and one free; carpels one; fruit a bean. The trees are usually spiny.

1   Flowers irregular (Sweetpea-like) **2**
1   Flowers regular (*Parkinsonia*-like) **7**
2   Leaves pinnately divided (feather-like) **3**
2   Leaves palmately divided (hand-like). **6**
3   Tree with branches bearing long spines. . . . . **Desert-Ironwood**
. . . . (*Olneya tesota*, p. 119)
3   Suffruticose or annual herbs; not spine bearing. **4**

4   Pods curved; flowers lavender . . . . **Sheep Milkvetch**
    . . . . (*Astragalus nothoxys*, p. 116)
4   Pods straight; flowers yellow 5
5   Flowers ¾ inch long; leaflets 1½ to 2 inches long; pods 1½ inch long
    . . . . **Broom Deerweed** (*Acmispon rigidus*, p. 54)
5   Flowers ¼ inch long; leaflets ¾ inch long; pods ½ inch long . . . .
    . . . . **Coastal Deerweed** (*Acmispon maritimus*, p. 53)
6   Leaflets broad, densely hairy; flowers light blue-lavender . . . .
    . . . . **Arizona Lupine** (*Lupinus arizonicus*, p. 117)
6   Leaflets linear, slender, sparsely hairy; flowers dark blue . . . .
    . . . . **Mojave Lupine** (*Lupinus sparsiflorus*, p. 118)
7   Annual or rarely biennial; flowers 1 inch across, orange-yellow . . . . .
    . . . . **Coves' Cassia** (*Senna covesii*, p. 58)
7   Trees; flowers 1 inch across or smaller; yellow or cream-color 8
8   Leaflets 1/16 inch long, far apart on the midrib; stems yellow-green,
    each spine-tipped; flowers ¾ to 1 inch across . . . .
    . . . . **Little-Leaf Palo-Verde** (*Parkinsonia microphylla*, p. 55)
8   Leaflets 3/16 inch long or longer, close together on the midrib; stems
    dark green, not spine tipped; flowers minute, many in crowded long
    spikes 9
9   Leaflets 3/16 inch long, spines on stems ⅛ inch long, curved like a
    cat's claw; flowers yellow. . . **Catclaw Acacia** (*Senegalia greggii*, p. 57)
9   Leaflets 3/16 to 5/16 inches long; spines 3/16 to ½ inches long,
    straight; flowers cream-yellow. . . . . **Velvet Mesquite**
    . . . . (*Prosopis velutina*, p. 94)

FOUQUIERIACEAE Ocotillo Family
   Spiny shrub, with several erect or slightly spreading stems from the base, bearing leaves but for a short time in the summer. The spines formed by the hardened midribs of the leaves of the previous seasons; sepals five, united; petals five, united; stamens ten; the stamens are joined to the corolla tube; fruit an ovoid capsule with many seeds.
   Ocotillo (*Fouquieria splendens*, p. 143)

GERANIACEAE Geranium Family
   Ours annual herbs; leaves lobed; flowers regular, in few flowered axillary pedunculate clusters; sepals five; petals five; stamens five; carpels five, united, the united styles forming a persistent column; fruit a capsule, each carpel breaking away from the column at maturity. In the genus Erodium the carpel tails are long-beaked and become spirally twisted.

1   Leaf blades pinnately divided; petals small, ¼ inch long or less, pale
    purplish . . . . **Red-Stem Stork's-Bill** (*Erodium cicutarium*, p. 121)
1   Leaf blades merely lobed; petals large, ⅜ inch long or more; purple .
    . . . . **Texas Stork's-Bill** (*Erodium texanum*, p. 122)

HELIOTROPIACEAE Heliotrope Family
Formerly placed in **Borage Family** (Boraginaceae).
Alkali Heliotrope (*Heliotropium curassavicum*, p. 123)

Former **HYDROPHYLLACEAE** Waterleaf Family
Ours annual herbs; leaves alternate, simple or compound; calyx of five more or less united sepals, the sinuses often somewhat appendaged; corolla of five united petals, the tube funnelform; stamens five; ovary superior; fruit a capsule, seeds few. Now treated as a sub-family of the **Borage Family** (Boraginaceae).

1 Leaves simple; flowers dark red-purple; plants small, stems procumbent . . . .**Fiddleleaf** (*Nama hispida*, p. 110)
1 Leaves with the margins deeply lobed; flowers blue; plants erect or with slender trailing stems. 2
2 Flowers numerous, borne on one side of the coiled stem in a helicoid cyme. . . . . . . **Notch-Leaf Scorpionweed** (*Phacelia crenulata*, p. 111)
2 Flowers few, solitary or far apart in the inflorescence. . . . .
. . . . (*Pholistoma auritum*, p. 112)

KRAMERIACEAE Krameria Family
Shrubs, branched throughout, the branches covered with a short silvery gray pubescence. Flowers irregular, sepals five, petal-like; petals five, reduced; stamens three; pistil simple.
White Ratany (*Krameria bicolor*, p. 124)

LAMIACEAE Mint Family
Shrubs or herbs, stems four-sided; leaves various; inflorescence terminal spikes; flowers irregular, calyx five-lobed, persistent; corolla five-toothed, bilabiate; stamens two; ovary deeply four-lobed; fruit four small nutlets in the persistent calyx.

1 Shrubs 4 to 8 feet high; leaves small, 1 inch long, light green; flowers light purple . . . .**Desert-Lavender** (*Condea emoryi*, p. 125)
1 Annual herb, 12 to 18 inches high; leaves 1½ to 2 inches long, dark green, rugose; flowers bluish. . . . .**California Sage**
. . . . (*Salvia columbariae*, p. 126)

LOASACEAE Loasa Family
Annuals or perennials; stems white; leaves simple, deeply covered with short coarse barbed hairs; hypantheum tubular; calyx lobes five; petals five to ten; stamens numerous; capsule slender.

1 Flowers large, 1½ to 2 inches across, petals ten, large, bright yellow .
. . . .**Long-Lobe Blazingstar** (*Mentzelia longiloba*, p. 60)
1 Flowers small, ¼ inch across, petals five 2
2 Flowers yellow; stems glossy white and very smooth . . . .
. . . .**White-Stem Blazingstar** (*Mentzelia albicaulis*, p. 59)

2   Flowers cream-white, the five stamens apparently on the outside of the petals; stems rough . . . . **Thurber's Sandpaper-Plant** . . . . (*Petalonyx thurberi*, p. 96)

MALPIGHIACEAE Barbados-Cherry Family
Low twining perennial with woody stems; leaves opposite, narrow, lanceolate, ½ to 1 inch long, pubescent; sepals five; petals five; stamens five; styles united; fruit a samara ½ inch long.
Slender Janusia (*Janusia gracilis*, p. 61)

MALVACEAE Mallow Family
Annual or perennial herbs; leaves variously dissected or lobed, alternate; plants with a stellate pubescence throughout; flowers showy; sepals five, united at the base and bearing three to twelve bracts; petals five, united at the base to the base of the stamen column; stamens united below into a conspicuous column which stands erect in the center of the flower, the anthers are borne at the top or down the sides of the column; pistil of five to many carpels; fruit a capsule. The flowers are similar to those of the common Hollyhock which is a member of the same family.

1   Flowers small, ¼ to ⅜ inches across, pink or white **2**
1   Flowers large, showy, 1½ inches across, orange or lavender **3**
2   Annual; leaves nearly round, thin, dark green, sparsely pubescent; flowers white, the petals tipped with lavender . . . . **Cheeseweed Mallow** (*Malva parviflora*, p. 97)
2   Shrub; with many slender gray branches; leaves small, heart-shaped; flowers pale pink, the petals with a darker pink line down the center . . . . **Pelotazo** (*Abutilon incanum*, p. 145)
3   Leaves gray-green, ovate; flowers lavender with dark purple centers; calyx bracts twelve . . . . . **Rock Hibiscus** (*Hibiscus denudatus*, p. 128)
3   Leaves yellow-green, shallowly lobed at the base; flowers orange or orange-red; calyx bracts three **4**
4   Winter annual 18 inches high; flowers yellow-orange . . . . **Coulter's Globemallow** (*Sphaeralcea coulteri*, p. 146)
4   Perennial 3 feet high; flowers orange or red-orange . . . . **Emory's Globemallow** (*Sphaeralcea emoryi*, p. 147)

NOTE The Mallow Family now includes **Sterculia Family** (Sterculiaceae), represented here by genus **Ayenia**. These are perennials with alternate simple leaves; five sepals; five petals; stamens joined into a tube; carpels five.
California Ayenia (*Ayenia compacta*, p. 127)

NYCTAGINACEAE Four-o'Clock Family
Annual or perennial herbs; leaves opposite, equal or very unequal; flowers regular, clustered and subtended by an involucre; calyx colored,

corolla-like; petals none; stamens inconspicuous except in the larger flowers; ovary one-celled.

1 Plants prostrate or nearly so; flowers lavender, purple, or red-purple 2
1 Plants erect or only partly procumbent and trailing 4
2 Flowers in large clusters resembling those of the cultivated Verbena; pink-lavender .... **Desert Sand Verbena** (*Abronia villosa*, p. 129)
2 Flowers purple or red-purple, apparently solitary 3
3 Branches 2 to 3 feet long, spreading; inflorescence leafless, branched; flowers 1/16 inch across, deep red-purple .... **Scarlet Spiderling** .... (*Boerhaavia coccinea*, p. 131)
3 Branches 12 to 20 inches long; inflorescence leafy; flowers axillary, ½ inch across, purple or lavender .... **Trailing Windmills** .... (*Allionia incarnata*, p. 130)
4 Plants trailing, stems very hairy; opposite leaves of equal size, large, heart-shaped, sticky, covered with glandular hairs; flowers ⅜ inch across, cream-colored .... **Desert Wishbonebush** .... (*Mirabilis laevis*, p. 98)
4 Plants erect; stems minutely hairy; opposite leaves very unequal, the larger one oblong, red above, green on the lower surface, not sticky; flowers 3/16 inch across, pale pink .... **Erect Spiderling** .... (*Boerhaavia erecta*, p. 132)

OLEACEAE Olive Family
Ours a low suffrutescent herb; leaves simple, entire; sepals five to six or wanting as in ours. Corolla rotate , the five to six petals united at the base; stamens two to three; ovary two-celled; stigma capitate; fruit circumscissle near the middle.
Rough Menodora (*Menodora scabra*, p. 62)

ONAGRACEAE Evening-Primrose Family
Ours annual herbs; leaves simple, alternate; sepals four, united at the base; petals four; stamens eight; pistil of four carpels; fruit a four-valved capsule which splits when ripe.

1 Plants 6 to 12 inches tall; leaves broad, margins irregularly toothed to pinnatifid at the base; flowers ½ inch across, cream colored .... **Browneyes** (*Chylismia claviformis*, p. 99)
1 Plants 18 to 24 inches tall; leaves very narrow, margins entire or inconspicuously toothed; flowers ¼ to ⅜ inches across, light red-orange .... **California-Primrose** (*Eulobus californicus*, p. 63)

PAPAVERACEAE Poppy Family
Ours an annual herb; leaves deeply dissected; flowers regular; sepals fused into a pointed cap which comes off when the flower is ready to open; petals four; stamens numerous; fruit a one-celled capsule.
California Poppy (*Eschscholzia californica*, p. 148)

PLANTAGINACEAE Plantain Family

Ours an annual herb, with numerous basal leaves; inflorescence spikate, flowers perfect, sepals four; corolla tubular with four spreading papery lobes; stamens two; fruit a circumscissle capsule; seeds few.

Desert Plantain (*Plantago ovata*, p. 100)

POACEAE Grass Family

Annuals or perennials; low growing; leaves slender, long, and with sheathing bases. The flowers are small, composed of two glumes surrounding a lemma and a palea. These structures are all more or less scale-like and overlapping one another. The pistil is small and bears two long feathery stigmas. The three stamens are usually quite conspicuous when the grass is in full bloom.

1   Spikes several, radiating umbrella-like from a single point at the apex of the flower stalk . . . . . . **Bermuda Grass** (*Cynodon dactylon*, p. 161)
1   Spikes numerous, arranged along the central flower stalk. **2**
2   Heads distinctly purple; flowers each with three awns **3**
2   Heads green; flowers with none or one awn **4**
3   Awns ¾ inch long, spreading . . . . **Purple Three-Awn** . . . . (*Aristida purpurea*, p. 159)
3   Awns 1/16 inch long, not spreading . . . . **Six-Weeks Grama** . . . . (*Bouteloua barbata*, p. 160)
4   Flowers with one erect awn; leaves broad, partly enclosing the head, at least when young . . . . **Wall Barley** (*Hordeum murinum*, p. 162)
4   Flowers without an awn; leaves slender, not enclosing the loosely arranged flower head . . . . **Bigelow's Bluegrass** (*Poa bigelovii*, p. 163)

POLEMONIACEAE Phlox Family

Ours annual herbs; leaves alternate or clustered, simple or compound; calyx five-lobed; corolla tubular or campanulate, five-lobed; stamens five; ovary three-celled; styles united; stigmas three; fruit a three-celled capsule; seeds few.

1   Leaves divided; mostly basal. . . . . . . **Rosy Gilia** (*Gilia sinuata*, p. 135)
1   Leaves simple, thread-like; not basal. **2**
2   Flowers few, ¾ inch long, cream-yellow . . . . **Desert-Trumpets** . . . . (*Linanthus bigelovii*, p. 101)
2   Flowers numerous, ½ inch long, blue (rarely white) . . . . **Desert Woollystar** (*Eriastrum diffusum*, p. 134)

POLYGONACEAE Buckwheat Family

Annuals or perennials, leaves mostly in basal rosettes. Flowers small, solitary or in clusters from a single involucre; perianth of six parts; stamens four to nine; fruit three-angled, usually winged.

1   Flowers not subtended by an involucre; leaves heavy, 6 to 12 inches long; inner sepals becoming enlarged, papery and red-brown in fruit . . . . **Canaigre Dock** (*Rumex hymenosepalus*, p. 164)
1   Flowers subtended by an involucre; leaves 2 to 3 inches long; inner sepals not becoming enlarged **2**
2   Teeth of the involucre bristle-tipped; flowers tiny, inconspicuous . . **3**
2   Teeth of the involucre not bristle-tipped; flowers small but very showy **4**
3   Plants with long slender internodes; flowers white . . . . **Brittle Spineflower** (*Chorizanthe brevicornu*, p. 102)
3   Plants densely spiny, stems not showing through the spines; flowers yellow . . . . **Devil's Spineflower** (*Chorizanthe rigida*, p. 65)
4   Shrubs; leaves narrow, small, many on the branches throughout the plant . . . . . . **California Buckwheat** (*Eriogonum fasciculatum*, p. 137)
4   Annuals; leaves broad, in basal rosettes **5**
5   Plants coarse, 2 to 3 feet high **6**
5   Plants delicate, 8 to 10 inches high . . . . . . **Thomas' Wild Buckwheat** . . . . (*Eriogonum thomasii*, p. 138
6   Stems inflated near the tips; leaves dark green; flowers yellow . . . . **Desert Trumpet** (*Eriogonum inflatum*, p. 66)
6   Stems not inflated; leaves densely felted with a gray wool; flowers pink . . . . **Flatcrown Buckwheat** (*Eriogonum deflexum*, p. 136)

PTERIDACEAE Maidenhair Fern Family
   Ferns, low growing rather inconspicuous plants found under rocks and in moist places. Flowers none, spores borne on the under surfaces of the fronds.

1   Fronds palmate or fan-like at the summit of the stipe . . . . **Copper Fern** (*Bommeria hispida*, p. 165)
1   Fronds pinnate or feather-like along the sides of the stipe . . . . **Pringle's Lip Fern** (*Myriopteris pringlei*, p. 166)

RANUNCULACEAE Crowfoot Family
   Ours an annual herb; calyx of five sepals which are petal-like, irregular, colored; the upper sepal is spurred; petals four, reduced to small erect appendages; stamens numerous; pistils several; seeds many.
   Ocean-Blue Larkspur (*Delphinium parishii*, p. 139)

RUBIACEAE Madder Family
   Ours a perennial herb, shrub-like; stems four-angled; leaves whorled, the stipules often as large as the leaves; flowers small, in axillary racemes. Sepals absent; corolla rotate, with four-lobes. Fruit of two nearly separate, one-seeded carpels, dry.
   Fendler's Bedstraw (*Galium fendleri*, p. 67)

SANTALACEAE Sandalwood Family

Evergreen plants parasitic on shrubs and trees; yellow-green; leaves scale-like, opposite; flowers small and inconspicuous, greenish, dioecious; sepals two to five; stamens of the same number as the sepals; ovary inferior, one-celled; fruit a berry with a viscid endocarp.

Mesquite Mistletoe (*Phoradendron californicum*, p. 167)

Former **SCROPHULARIACEAE** Figwort Family

Ours annual herbs; leaves alternate or opposite, simple and entire or lobed; flowers irregular, calyx of four to five more or less united sepals; corolla of five united petals, bilabiate; stamens four to five, usually of two kinds; ovary two-celled; fruit a two-celled capsule; seeds numerous.

1   Leaves opposite, dark green, ovate, shallowly toothed on the margins; flowers yellow, 1 inch long; now placed in Phrymaceae . . . . **Common Yellow Monkey-Flower** (*Erythranthe guttata*, p. 64)
1   Leaves alternate, medium green, deeply three-lobed; flowers purple-pink, ¾ inch long, upper petals bent forward like a hood; now placed in Orobanchaceae . . . . **Purple Indian-Paintbrush** . . . . (*Castilleja exserta*, p. 133)

SELAGINELLACEAE Selaginella Family

Low growing moss-like plants. Leaves scale-like, about ⅛ inch long. Spores borne on small erect "cones." These cones are similar to the vegetative branches in color but are four-sided, the minute orange spores being borne in the axils of the leaves.

Arizona Spikemoss (*Selaginella arizonica*, p. 168)

SOLANACEAE Nightshade Family

Herbs or shrubs; leaves alternate, solitary or fascicled; flowers regular, calyx of five more or less united sepals; corolla of five united petals, the tube elongated, tubular or funnelform, or short; stamens five, inserted on the corolla tube; ovary two-celled; fruit a berry or capsule, seeds many.

1   Flowers purple or lavender 2
1   Flowers cream color or yellow-green 3
2   Plants 12 to 18 inches tall, rather coarse, stems and leaves with prickles; flowers 1 inch across, saucer-shape . **Silverleaf Nightshade** . . . . (*Solanum elaeagnifolium*, p. 141)
2   Plants spreading, 2 to 3 inches tall, slender; leaves linear, smooth; flowers ¼ inch across, pale lavender . . . . **Wild Petunia** . . . . (*Calibrachoa parviflora*, p. 140)
3   Thick spiny shrub; leaves small, fascicled; flowers pale cream, ⅜ inch long . . . . **Red-Berry Desert-Thorn** (*Lycium andersonii*, p. 103)
3   Coarse annuals or perennials, not spiny; leaves large, solitary; flowers yellow-green, tubular, 1 to 2 inches long 4

4   Plants 18 to 24 inches tall; stems and leaves short hairy . . . .
. . . . **Desert Tobacco** (*Nicotiana obtusifolia*, p. 69)
4   Plants 4 to 8 feet tall; stems and leaves glossy green, mostly covered
with a light bloom . . . . **Tree-Tobacco** (*Nicotiana glauca*, p. 68)

ZYGOPHYLLACEAE Caltrop Family
    Herbs or shrubs, leaves compound; sepals five; petals five; stamens ten;
pistil of five united carpels.

1   Shrub, 3 to 8 feet high; fruits covered with short soft white hairs . . . .
. . . . **Creosotebush** (*Larrea tridentata*, p. 71)
1   Herbs, erect or procumbent; leaves with several pairs of leaflets. . . . 2
2   Flowers ⅜ inch across, yellow; fruits with several long, sharp spines
. . . . **Puncture-Vine** (*Tribulus terrestris*, p. 72)
2   Flowers 1 inch across, orange; fruits not spiny . . . .
. . . . **Arizona Poppy** (*Kallstroemia grandiflora*, p. 70)

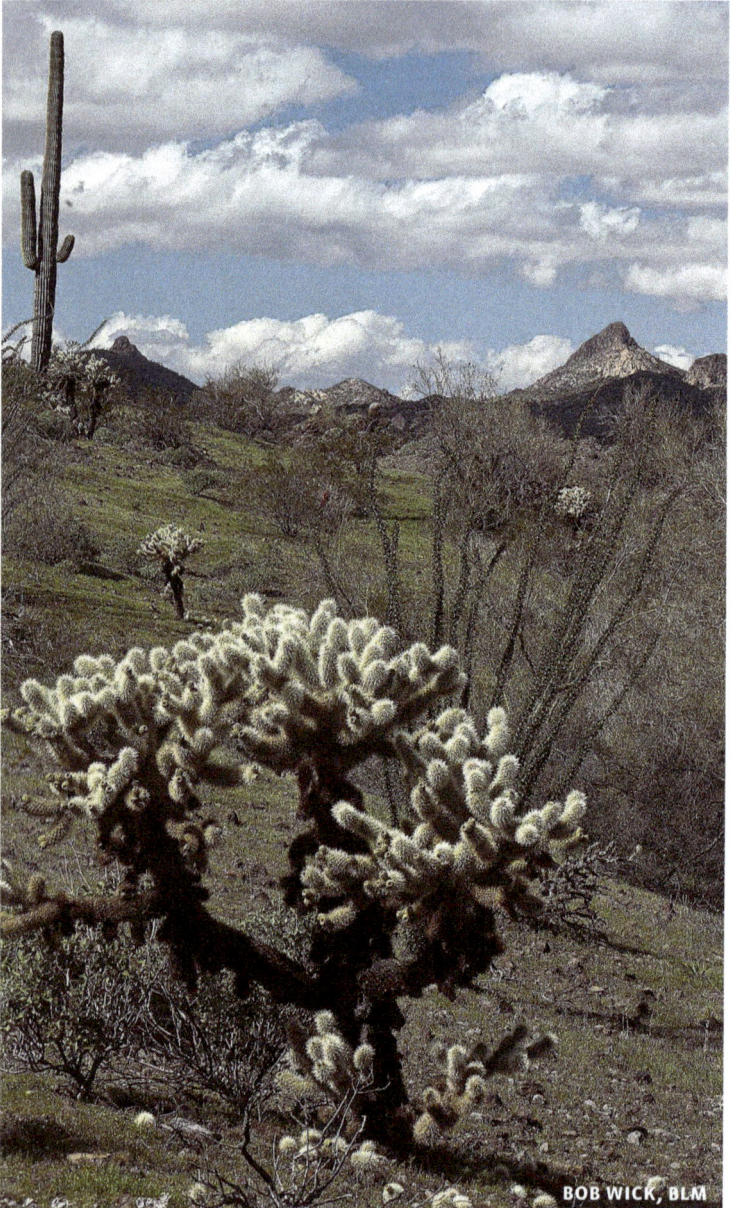

Teddy-Bear Cholla (*Cylindropuntia bigelovii*) with Ocotillo (*Fouquieria splendens*) and Saguaro (*Carnegiea gigantea*). Hummingbird Springs Wilderness, Maricopa County, Arizona.

ERIC KNIGHT

AMARANTHACEAE Amaranth Family

WOOLLY HONEYSWEET

Tidestromia lanuginosa (Nutt.) Standl.

DESCRIPTION Prostrate annual herb, with many **stems** from the base, these branched throughout, 12 to 15 inches long, light green to red. **Leaves** opposite or sometimes three in a whorl, ¾ to 1½ inch long, ½ to ¾ inch wide, the blade ovate and tapered to the slender long petiole. Leaves and stems covered so densely with short stellate or plumose hairs as to appear white. **Flowers** ⅛ inch long, axillary, usually two in a cluster, sessile; bracts papery, hairy tipped, green-yellow in color; sepals erect, oblong, pale yellow, sparsely stellate hairy. **Fruit** an ovoid indehiscent utricle.

FLOWERING Summer to fall.

ELEVATION Up to 5,500 feet.

WHERE FOUND Common on sandy open soils.

NOTE The whitish mats of this plant are conspicuous soon after summer rains, and are useful in slowing the blowing of sandy soils.

MATT BERGER

ASTERACEAE Aster Family

PARISH'S GOLDENEYE

Bahiopsis parishii (Greene) E.E.Schill. & Panero

SYNONYM *Viguiera deltoidea* var. *parishii* (Greene) Vasey & Rose

DESCRIPTION Perennial, 18 to 24 inches tall, somewhat branched throughout, branches nearly white, covered with a short rough pubescence.
Leaves mostly opposite, petiole ⅛ inch long, blade ¾ to 1 inch long,
ovate-triangular, margins irregularly dentate; the leaves are dark green,
covered with a short rough pubescence and have three prominent veins
from the base. **Flower heads** about 1¼ inches across, **ray flowers** about
ten, ½ inch long, bright yellow, **disk flowers** dark yellow.

FLOWERING Summer to fall.

ELEVATION Up to 3,500 feet.

WHERE FOUND Dry mesas, rocky slopes.

RACHEL STRINGHAM

ASTERACEAE Aster Family

DESERT MARIGOLD

Baileya multiradiata Harv. & A. Gray

DESCRIPTION Biennial or perennial, 18 to 24 inches high, many **stems** from or near the base. At the base, a cluster of alternate, pinnatifid, densely white woolly leaves. **Flower heads** borne on long slender leafless stems. The flower heads are 1½ to 2 inches across, the disks are bright yellow as are also the several rows of showy, bright persistent rays which become reflexed in age.

FLOWERING Spring to summer.

ELEVATION Up to 5,000 feet.

WHERE FOUND Lower deserts, open sunny areas, sandy plains and mesas, gravelly washes, hillsides, dry soils, roadsides. Common in southern Arizona but not abundant around Phoenix.

NOTE Desert marigold is often used as a landscape plant by homeowners and used as a roadside planting by state highway departments. It is stated that horses crop the flower heads, but fatal poisoning of sheep and goats eating this plant on overgrazed ranges has been reported.

ASTERACEAE Aster Family

BRITTLEBUSH, GOLDEN HILLS

Encelia farinosa A. Gray ex Torr.

DESCRIPTION A conspicuous shrub of the desert landscape, blooming in early spring and covering the hillsides with a blanket of yellow. The plants are 1½ to 3 feet high, forming very large rounded clumps, have gray-green **leaves** and numerous leafless flower stalks rising 6 to 12 inches above the body of the plant. Each flower stalk bears a golden-yellow **flower head** 1½ inch across.

FLOWERING Spring.

ELEVATION Up to 3,000 feet.

WHERE FOUND Common at low elevations with creosotebush, and on dry rocky slopes and hillsides.

NOTE The stems exude a gum which was chewed by the Indians; used as incense in the churches of Baja California.

NEBROOKS

ASTERACEAE Aster Family

COMMON SUNFLOWER
Helianthus annuus L.

DESCRIPTION Coarse annual, 4 to 6 feet tall, branched throughout. **Leaves** 4 to 8 inches long, the blades 1½ to 3 inches wide, ovate-acute, dark green. Stems and leaves rough hirsute and with a strong resinous odor. **Flower heads** numerous, showy, 2 to 3 inches across, rays about 15, ¾ to 1 inches long, bright yellow; disk 1 inch across, dark brown.

FLOWERING Late summer to fall.

ELEVATION Up to 7,000 feet.

WHERE FOUND Common in a variety of habitats such as roadsides, irrigated fields, disturbed areas, etc.

COREY FARWELL

ASTERACEAE Aster Family

CAMPHORWEED

Heterotheca subaxillaris (Lam.) Britton & Rusby

DESCRIPTION Annual or biennial, 3 to 6 feet high, one heavy stem from the base, this with many branches from near the top, each branch bearing several flower heads. **Leaves** 1 to 2½ inches long, oblong-ovate, entire or toothed, clasping at the base. Leaves and stems very hairy and with a more or less sticky feeling from the numerous glandular hairs. **Flower heads** numerous, very showy, 1 inch across, rays about 25, ⅜ inch long, $1/16$ inch wide, bright yellow; disk 3/16 inch across, bright yellow.

FLOWERING Summer to fall.

ELEVATION Up to 5,500 feet.

WHERE FOUND Very common along roadsides especially; disturbed areas, dry desert washes, ditches.

NOTE Known as "camphorweed" because of the odor of the plant.

ROSIE STEINBERG

ASTERACEAE Aster Family

SOUTHERN GOLDENBUSH

Isocoma pluriflora Greene

DESCRIPTION Shrub, 2 to 3 feet high, many branches from the base, these brittle, erect or nearly so and not branched, tan-yellow and sparsely puberulent. **Leaves** alternate, 1 to 2 inches long, ⅛ inches wide, linear-oblanceolate, yellow-green, margins scarious, surfaces inconspicuously scabrous. The plant has a heavy resinous odor. **Flower heads** numerous in a flat topped cluster, several clusters at the tip of each stem. Flower heads ⅜ inch long, involucral bracts in several rows, overlapping, pale green with dark green thickened spices. **Flowers** all disk, yellow; pappus of numerous cream-colored inconspicuously plumose bristles.

FLOWERING Summer.

ELEVATION Up to 5,000 feet.

WHERE FOUND Often in creosotebush and mesquite communities. mesas and plains with saline, sandy, clay or alkaline soils, common along roadsides and on canal banks.

NOTE Also called "Jimmyweed." The plant often occupies overgrazed range land and is a common roadside weed in the irrigated districts. Plants are generally unpalatable, but when eaten in quantity by cattle cause the disease known as "milk sickness," or "trembles," which is transmissible through the milk to humans.

GEORGE WILLIAMS

ASTERACEAE Aster Family

GOLDFIELDS

Lasthenia gracilis (DC.) Greene

SYNONYM *Baeria gracilis* DC.

DESCRIPTION Very slender delicate annual, 3 to 6 inches tall, usually a solitary stem and flower head. **Leaves** ½ to ⅝ inches long, narrowly linear, pale green. Stems and leaves very sparsely hairy. **Flower heads** ½ to ⅝ inches across, rays about twelve, 3/16 inches long, light yellow; disk ¼ inches across, medium yellow.

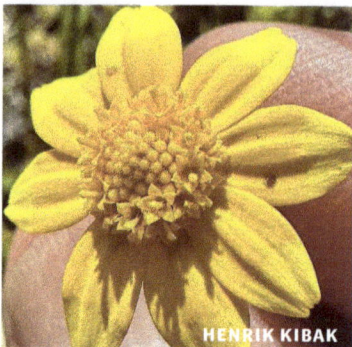

HENRIK KIBAK

FLOWERING Spring.

ELEVATION Up to 4,500 feet.

WHERE FOUND Forming patches in open areas, mesas, and plains; often where soils are alkaline and clayey.

NOTE In spring extensive areas are sometimes carpeted with the bright yellow flowers of this plant, which is reported to be cropped by horses.

SUE CARNAHAN

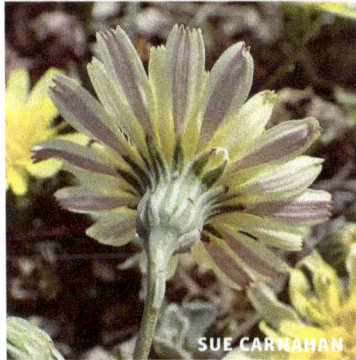

SUE CARNAHAN

ASTERACEAE Aster Family

DESERT DANDELION
Malacothrix fendleri A. Gray

DESCRIPTION Many slender flower stalks from the base, 6 to 8 inches high. **Leaves** basal, 1 to 2 inches long, slender and deeply pinnately dissected and finely toothed. **Flower heads** 1 to 1¼ inches across, the outer rays long, the center ones short, all bright yellow.

FLOWERING Spring.

ELEVATION 2,000 to 5,000 feet.

WHERE FOUND Common on sandy plains, mesas, rocky slopes.

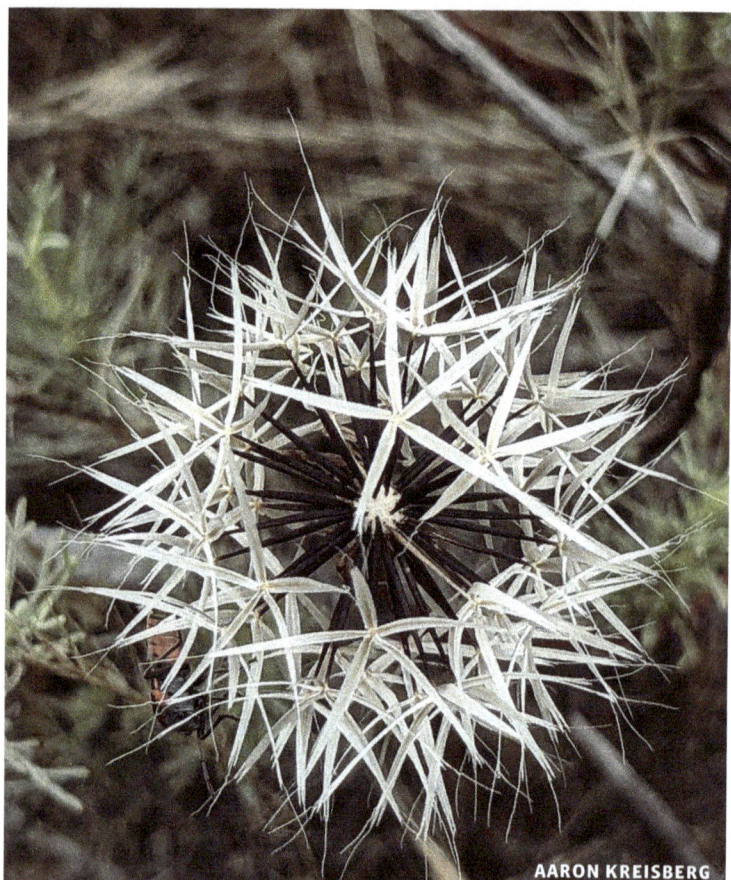

AARON KREISBERG

ASTERACEAE Aster Family

SILVER PUFFS
Microseris lindleyi (DC.) A. Gray

SYNONYM *Uropappus lindleyi* (DC.) Nutt.

DESCRIPTION Several **stems** from the base, 12 to 15 inches high, these smooth and hollow. **Leaves** 1½ to 2 inches long, narrow, slender, and with several short narrow lobes on either side. **Flower heads** small, not showy, yellow. After the head has gone to seed it is very pretty and conspicuous: it is about 1½ inch across and each seed is tipped with the silver, papery, star-like pappus; the involucral bracts are quite long.

FLOWERING Spring.

ELEVATION Up to 5,000 feet.

WHERE FOUND Common in various habitats: mesas, plains, grasslands, chaparral, deserts and sandy desert flats; prefers well-drained soils, rocky sites, disturbed areas and roadsides.

RACHEL STRINGHAM

ASTERACEAE Aster Family

LEMMON'S RAGWORT
Senecio lemmonii A. Gray

DESCRIPTION Annual, branched throughout, 1 to 2½ feet high. Stems slender, glabrous. **Leaves** few, perfoliate, 2 to 3½ inch long, ¾ to 1 inches wide, entire or shallowly toothed, especially near the base, thin, medium green. **Flower heads** in clusters at the tips of the stems. **Heads** 1 inch across, the bright yellow rays ⅜ inch long, the disk ⅜ inch across, yellow. Involucre of numerous slender green bracts.

FLOWERING Spring.

ELEVATION Up to 3,500 feet.

WHERE FOUND Common on rocky slopes and dry hillsides, usually among other shrubs.

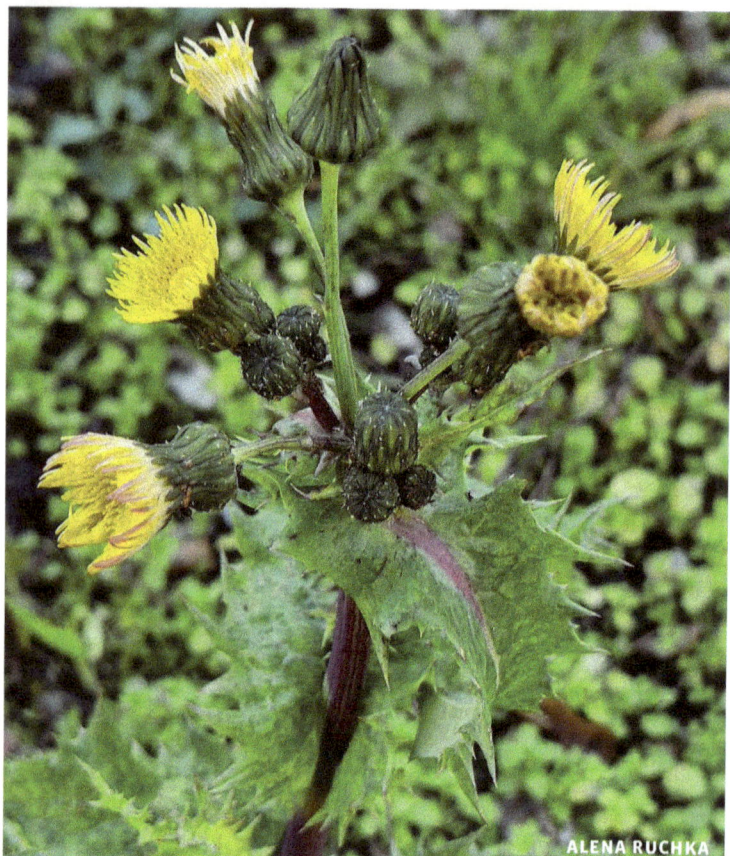

ALENA RUCHKA

ASTERACEAE Aster Family

PRICKLY SOW-THISTLE
Sonchus asper (L.) Hill

DESCRIPTION Introduced annuals, 1 to 3 feet high, branched somewhat, the branches finely angled and pale yellow-green. When broken the tissues give off a milky juice. The upper parts of the stems and the involucres of the flowers bear numerous glandular hairs. The **leaves** are 3 to 5 inches long, pinnately lobed, the lobes being irregular and toothed. At the base of the plant the leaves are broadened and clasp the stems. **Flower heads** numerous, 1 to 1¼ inches across, light yellow and composed of ray flowers only. When in seed the heads form the familiar downy white 'puff balls.'

FLOWERING Spring to summer.

ELEVATION Up to about 8,000 feet.

WHERE FOUND Roadsides, yards, gardens, fields, along streams and disturbed areas.

SUE CARNAHAN

ASTERACEAE Aster Family

YELLOWDOME

Trichoptilium incisum (A. Gray) A. Gray

DESCRIPTION Annual, 2 to 4 inches tall, branched from the base. **Leaves** numerous, crowded near the base of the plant, ½ to 1 inches long, slender at the base, broadened and deeply toothed above. The leaves and lower stems are densely white woolly. Peduncles slender, leafless, short glandular hairy, rising 1 inches to 2 inches above the leaves. The slender involucral bracts are sparsely long white hairy. **Flower heads** to ½ inches across, numerous, pale yellow. Pappus of white hairs.

FLOWERING Summer to fall.

ELEVATION Up to about 2,500 feet.

WHERE FOUND Sandy or gravelly soils, dry slopes, mesas and plains; often with creosotebush; not common.

ANDREW MEEDS

ASTERACEAE Aster Family

AMERICAN THREEFOLD

Trixis californica Kellogg

DESCRIPTION Shrub, 2 to 3 feet high, branched throughout, stems slender and brittle. **Leaves** 1 to 2 inches long, ovate-lanceolate, the margins shallowly toothed, the surfaces a glossy green. **Flower heads** many, ¾ inch long, nearly 1 inch across, of disk flowers only. Individual flowers ¾ inch long, bright yellow.

FLOWERING Spring.

ELEVATION Up to 5,000 feet.

WHERE FOUND Common at lower to middle elevations; southern exposures, dry sunny areas, ridges, rocky hillsides and slopes, desert washes.

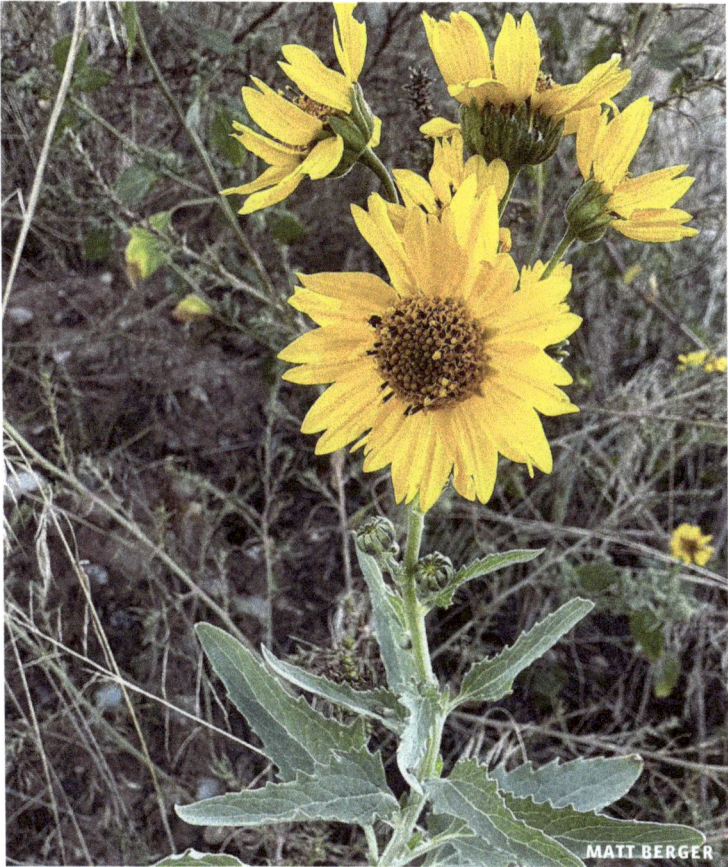

MATT BERGER

ASTERACEAE Aster Family

COWPEN DAISY, GOLDEN CROWNBEARD

Verbesina encelioides (Cav.) Benth. & Hook. f. ex A. Gray

DESCRIPTION Coarse annual, 12 to 15 inches high, few stems from the base. **Leaves** 1½ to 2 inches long, ¾ to 1 inches wide, oblong, margins irregularly, coarsely toothed to nearly entire, light silver-green, and covered with fine appressed silky hairs especially on the lower surface. **Flower heads** 1¼ to 1½ inches across, rays ½ inch long, toothed at the blunt tips, bright yellow; disk ½ to ⅝ inches across, dark yellow.

FLOWERING Summer.

ELEVATION Up to 8,000 feet (usually much lower).

WHERE FOUND Roadsides, disturbed areas, fields; sandy, gravelly, silty or rocky soils; not common.

NOTE This plant is said to have been used by Indians and early settlers for boils and skin diseases. The Hopis were reported to bathe in water in which this plant has been soaked, to relieve the pain of spider bite.

RACHEL STRINGHAM

BORAGINACEAE Borage Family

RANCHER'S FIDDLENECK

Amsinckia intermedia Fisch. & C.A. Mey.

DESCRIPTION Coarse, 12 to 24 inches high. **Leaves** 2 to 3½ inches long, ½ inch wide, entire, acute, lanceolate, dark green. Stems and leaves coarsely hispid. **Inflorescence** a helicoid cyme. **Flowers** ⅜ inch long, corolla tube tubular, pale yellow at the base, orange-yellow above, corolla lobes rounded, spreading, orange-yellow.

FLOWERING Spring.

ELEVATION Up to 5,500 feet.

WHERE FOUND Common in dry open places, disturbed areas, sandy or gravelly soils.

SUE CARNAHAN

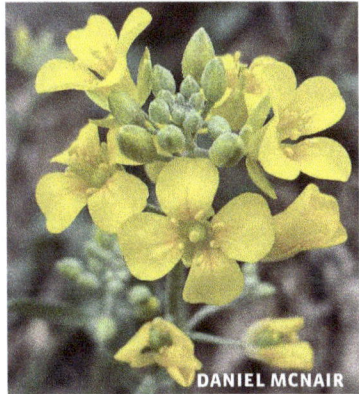

DANIEL MCNAIR

BRASSICACEAE Mustard Family

BLADDER POD

Physaria gordonii (A. Gray) O'Kane & Al-Shehbaz

DESCRIPTION Central **stems** erect, 6 to 10 inches high, the outer ones more or less procumbent with the ends erect. **Leaves** silvery green, oblanceolate, ½ to 1¼ inches long. **Flowers** bright yellow, ⅜ inch across, very showy, many out at the same time. **Pods** globose, 3/16 inch long, erect on the somewhat recurved pedicels. The stems and leaves are minutely stellate hairy.

FLOWERING Spring.

ELEVATION Up to 5,000 feet.

WHERE FOUND Very common on dry plains and mesas; sandy soils.

NOTE Extensive desert areas are colored in spring with the bright yellow flowers of this plant. It is reported to afford good forage, probably after the pods mature.

AUGUSTIN SOULARD

BRASSICACEAE Mustard Family

LONDON ROCKET

Sisymbrium irio L.

DESCRIPTION One main **stem** from the root, 8 to 24 inches high, this usually branched, each branch bearing flowers. **Leaves** mostly basal, 2 to 4 inches long, deeply dissected along the sides, light green. **Flowers** in a long raceme, light yellow, ⅛ to ³/₁₆ inches across. **Pods** 1 to 2 inches long, very slender. Seeds numerous.

FLOWERING Early spring.

ELEVATION Up to 4,500 feet.

WHERE FOUND Common weed of irrigated areas, open fields, disturbed places.

NOTE Introduced, and one of the troublesome weeds in Phoenix and the vicinity. It also harbors the False Chinch Bug which is so destructive to gardens.

CACTACEAE Cactus Family

BUCKHORN CHOLLA
Cylindropuntia acanthocarpa (Engelm. & J.M. Bigelow) F.M. Knuth

SYNONYM *Opuntia acanthocarpa* Engelm. & Bigel.

DESCRIPTION Much branched, 3 to 6 feet high, branches becoming woody. Branches alternate, making a narrow angle with the trunk; joints 1½ to 3 inches long, usually greatly exceeding this length, strongly tuberculate, the tubercles flattened, elongated and quite high. **Spines** 8 to 25 in a cluster, acicular, dark brown, covered with thin and lighter colored sheaths; spines ¾ to 1 inches long. **Flowers** large, red to yellow, 2 inches long and when fully open nearly as broad. The ovary with few prominent tubercles. **Fruit** dry, about 1¼ inches long, seeds ¼ to ⅜ inches broad, angular.

FLOWERING Spring.

ELEVATION Up to about 3,500 feet.

WHERE FOUND Common on sandy or gravelly soils.

NOTE The Pima Indians use the flower buds for food. The product, which is prepared by a steaming process, kept well and was eaten as needed, usually in combination with pinole or saltbush greens.

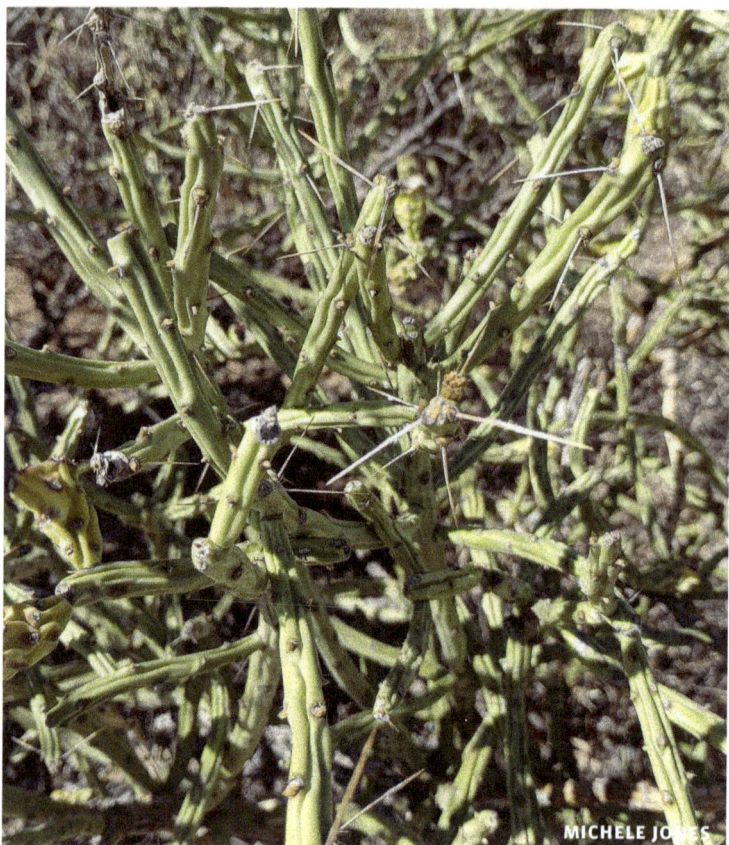

MICHELE JONES

CACTACEAE Cactus Family

ARIZONA PENCIL CHOLLA

Cylindropuntia arbuscula (Engelm.) F.M. Knuth

SYNONYM *Opuntia arbuscula* Engelm.

DESCRIPTION Shrub 6 to 8 feet high, often with a rounded, very compact top with numerous branches. Joints 2 to 3 inches long, ½ inch in diameter, with low distinct tubercles; **spines** one or rarely several, with loose straw colored sheaths. **Flowers** greenish yellow tinged with red, 1¼ inches long.

FLOWERING Spring.

ELEVATION Up to 3,000 feet.

WHERE FOUND Uncommon on sandy, gravelly washes in the desert.

NOTE Typically an tree-like shrub with compactly branched crown and well-developed trunk, but often the plant is less than 3 feet high and openly branched. The Papago Indians utilized the young joints of the pencil cholla and similar species as a boiled vegetable, but probably only in times of want.

BOB WALKER

CACTACEAE Cactus Family

TEDDY-BEAR CHOLLA
Cylindropuntia bigelovii (Engelm.) F.M. Knuth

SYNONYM *Opuntia bigelovii* Engelm.

DESCRIPTION Usually with a central, erect trunk 3 feet high or less, with short lateral branches, the upper ones erect. Joints 2 to 5 inches long, very stout, with closely set areoles and almost impenetrable armament of yellow **spines**. Tubercles slightly elevated, pale green, somewhat four-sided; spines and papery sheaths pale yellow. **Flowers** 1½ inches long, including the ovary, pale magenta to crimson or green-yellow.

FLOWERING Spring.

ELEVATION Up to 3,000 feet.

WHERE FOUND Occasional on sandy flats, gravelly to rocky washes, bajadas and hillsides; found in both Mojave and Sonoran Deserts.

NOTE Also known as Jumping Cholla; the combination of barbed spines and densely armed, very easily detached joints has earned a profound respect for this formidable cholla.

HARDY SMALLFEAT

CACTACEAE Cactus Family

CHRISTMAS CACTUS
Cylindropuntia leptocaulis (DC.) F.M. Knuth

SYNONYM *Opuntia leptocaulis* DC.

DESCRIPTION Usually bushy, often compact, 1 to 6 feet high, but sometimes with a short, definite trunk 2 to 4 inches in diameter, dull green with darker blotches below the areoles. The branches are slender, cylindric, ascending, tuberculate; the joints, especially the fruiting ones, are thickly set with short, usually spineless joints spreading at right angles to the main branches. These small joints are very easily detached and only ¾ to 2 inches long or less. **Spines** two to three together, ¾ to 2 inches long or less, sheaths closely fitting; areoles with short white wool. **Flowers** greenish or yellowish, ¾ to 1 inches long. **Fruit** small, globular to obovate or even clavate, red, to 1¼ inches long.

FLOWERING Spring.

ELEVATION Up to 5,000 feet.

WHERE FOUND Occasional in a number of habitats including grasslands, chaparral and oak-juniper woodlands, flats, and bajadas; prefers sandy loamy and gravelly soils.

NOTE Variable, especially in respect to development of the spines, but the small scarlet fruits, very slender stems, and relatively small size of the plant distinguish this species.

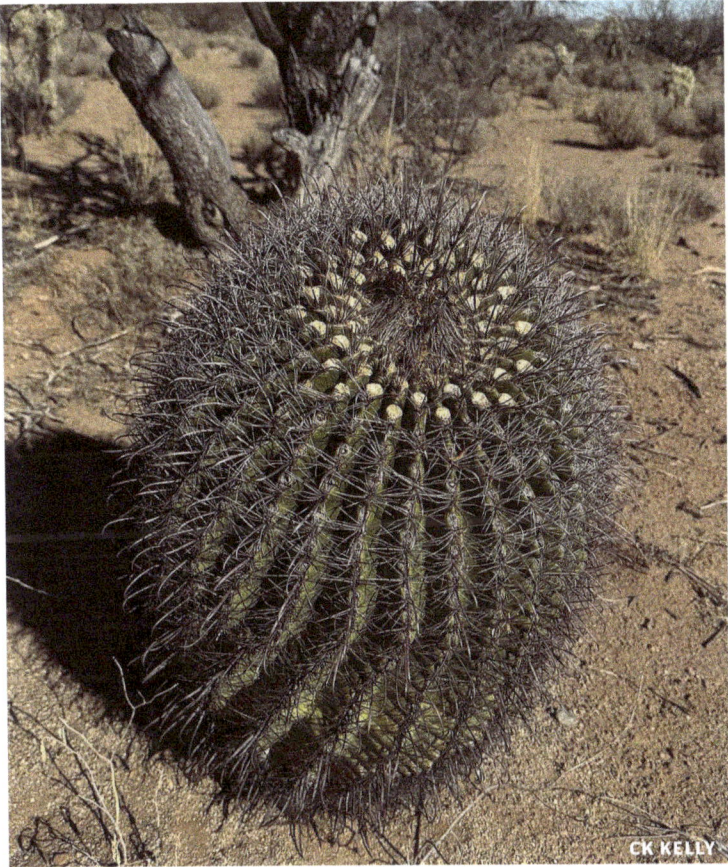

CK KELLY

CACTACEAE Cactus Family

BISNAGA, FISHHOOK BARREL CACTUS

Ferocactus wislizeni (Engelm.) Britton & Rose

SYNONYM *Echinocactus wislizeni* Engelm.

When young globular but becoming cylindric and much elongated when very old. Plants grow to be 6 feet high or more, one of the largest barrel cacti. Ribs numerous, often twenty-five, 1¼ inches high; areoles large, 1 inch long, brown felted. **Spines**: the radials thread-like to acicular, the centrals red to white, annular, one stouter than the others, usually flattened, strongly hooked and as long as 3 inches. **Flowers** yellow (sometimes red), 2 to 2½ inches long. **Fruit** yellow, oblong, scaly, 1½ to 2 inches long, seeds dull black.

FLOWERING Spring.

ELEVATION Up to 5,500 feet.

WHERE FOUND Fairly common in desert scrub, flats, bajadas, rocky areas, mountainsides, and grasslands.

CK KELLY

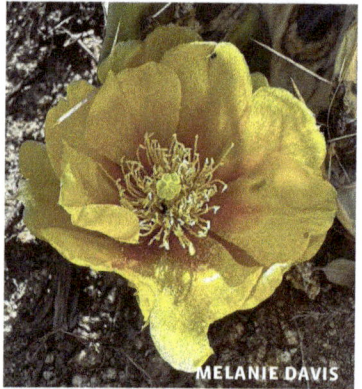

MELANIE DAVIS

CACTACEAE Cactus Family

BROWN-SPINED PRICKLY-PEAR
Opuntia phaeacantha Engelm.

DESCRIPTION Low, usually nearly prostrate, some branches ascending. **Joints** usually longer than broad, 4 to 6 inches long; areoles rather far apart, not heavy, the lower ones often spineless. **Spines** one to four, those on the sides of the joints more or less reflexed, brown, sometimes darker at the base, or nearly white throughout, the larger ones 2 to 2½ inches long. **Flowers** yellow, 2 inches across; ovary short. **Fruit** to 1½ inches long, much constricted at the base.

FLOWERING Spring.

ELEVATION To 7,500 feet.

WHERE FOUND Common throughout the Phoenix area.

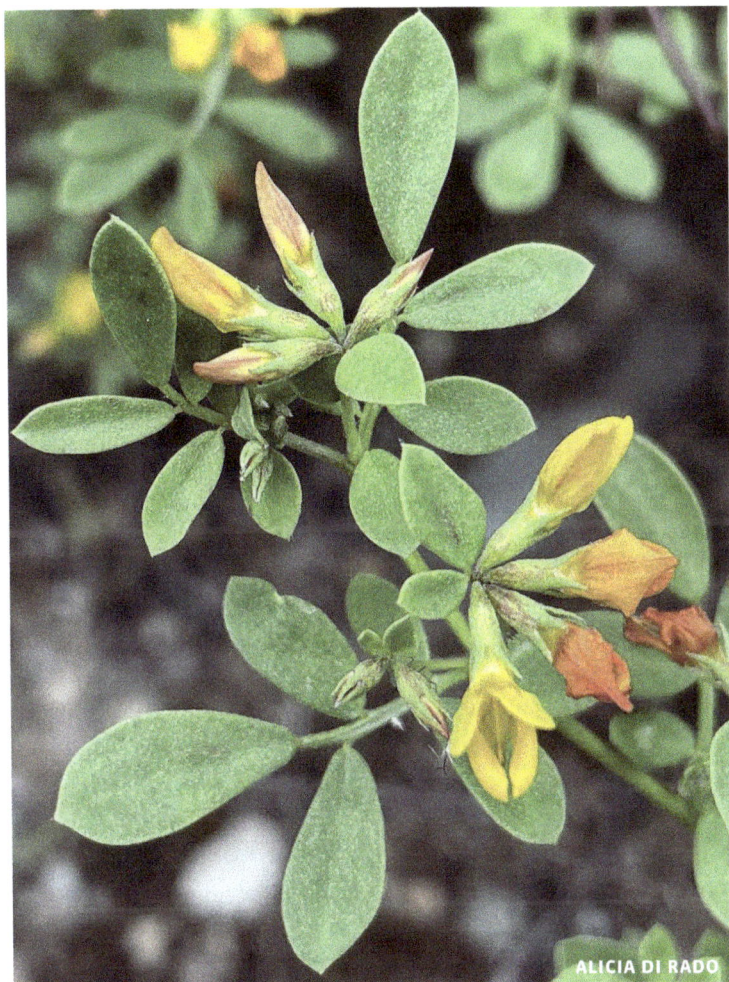

ALICIA DI RADO

FABACEAE Pea Family

COASTAL DEERWEED
Acmispon maritimus (Nutt.) D.D. Sokoloff

SYNONYM *Lotus humilis* Greene

DESCRIPTION Annual, about 4 inches tall, the outer branches procumbent.
Leaves 1 inches long, two to three pairs of leaflets, these ⅜ inch long,
¼ inch wide, obovate, dark green, very sparsely hairy. **Flowers** ¼ inches
long, yellow, the banner brown. **Pods** a narrow cylinder ½ inch long,
brown, glabrous.

FLOWERING Spring.

ELEVATION Up to 3,000 feet.

WHERE FOUND Common on dry hills and mesas.

RACHEL STRINGHAM

DON LOARIE

FABACEAE Pea Family

BROOM DEERWEED

Acmispon rigidus (Benth.) Brouillet

SYNONYM *Lotus argensis* Coville

DESCRIPTION Loosely branched throughout, **stems** slender, dark green, 12 to 15 inches long. **Leaves** ¾ inch long, one to two pairs of leaflets, these ½ to ¾ inches long, ⅛ inches wide, linear-oblanceolate, strigose. **Flowers** few, ¾ inch long, orange-yellow, the banner darker. **Pods** 1½ to 2 inches long, ⅛ inch wide, brown, glabrous.

FLOWERING Spring.

ELEVATION Up to 4,000 feet.

WHERE FOUND Uncommon on dry rocky slopes.

ABRAHAM ROMERO

FABACEAE Pea Family

LITTLE-LEAF PALO-VERDE
Parkinsonia microphylla Torr.

DESCRIPTION Tree 6 to 12 feet high, branched throughout, the branches yellow-green, and each tipped with a stout spine. **Leaves** ½ to ¾ inches long, leaflets four to six pairs, $^1/_{16}$ inch long, obovate, yellow-green. **Flowers** many, 1 inch across, bright yellow, the petals falling off easily.

FLOWERING Spring.

ELEVATION Up to 4,000 feet.

WHERE FOUND Common on dry rocky hillsides and mesas, less common in washes than blue palo-verde (*Parkinsonia florida*) which requires more regular water.

LITTLE-LEAF PALO-VERDE

CK KELLY

SUE CARNAHAN

FABACEAE Pea Family

CATCLAW ACACIA
Senegalia greggii (A. Gray) Britton & Rose

SYNONYM *Acacia greggii* A. Gray

DESCRIPTION Shrub 4 to 6 feet high, branched throughout, the **stems** slender and gray-brown. **Leaves** ¾ to 1 inches long, leaflets three to four pairs, ³/16 inch long, oblong-obovate, medium green. At the base of each leaf cluster is a short, brown, hooked spine. **Flowers** small, yellow, in a dense elongated spike, the stamens distinct and showy. **Pods** flat, about ½ inch wide (or more) and somewhat curved.

FLOWERING Spring.

ELEVATION Up to 4,500 feet.

WHERE FOUND Common in shrublands, arroyos, along streams, streambanks and washes; soils sandy, rocky, well-drained, often in alkaline soils.

NOTE The new foliage is relished by cattle in early spring, otherwise catclaw is valuable chiefly as a reserve food in times of drought or on depleted ranges. The Arizona Indians made meal of the pods, using it as mush and cakes. The flowers are one of the most important sources of honey for bees kept on the desert. The wood is very strong and was used locally for making doubletrees and singletrees, as well as for firewood. This is probably the most heartily disliked plant in the State, the sharp strong prickles tearing the clothes and lacerating the skin.

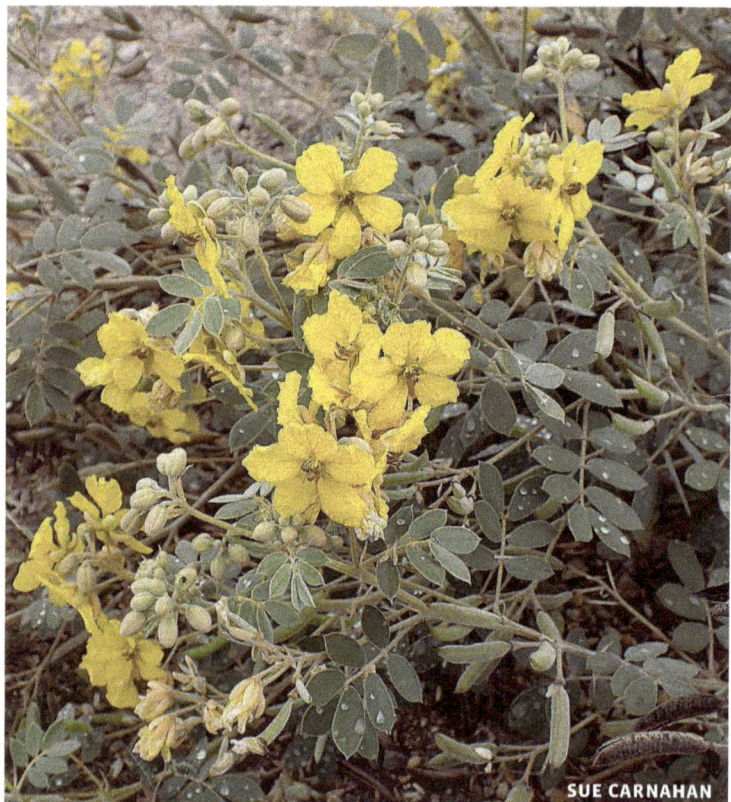

SUE CARNAHAN

FABACEAE Pea Family

COVES' CASSIA

Senna covesii (A. Gray) H.S. Irwin & Barneby

SYNONYM *Cassia covesii* A. Gray

DESCRIPTION Annual or biennial, woody at base, 1 to 2 feet high, somewhat branched. **Leaves** 1 to 3 inches long, leaflets four to six pairs, ½ inch long or more, obovate, pale green, with a close, dense pubescence. **Flowers** numerous, clustered at the ends of slender peduncles 1½ to 2 inches long. Flowers 1 to 1¼ inches across, bright orange-yellow, the petals veined in brown. **Pods** 1 inch long, dark brown, densely short-pubescent.

FLOWERING Spring to fall.

ELEVATION Up to 3,000 feet.

WHERE FOUND Uncommon on dry rocky slopes, sandy desert washes and mesas.

MY-LAN LE

LOASACEAE Blazingstar Family
WHITE-STEM BLAZINGSTAR
Mentzelia albicaulis (Douglas ex Hook.) Douglas ex Torr. & A. Gray

DESCRIPTION Branched throughout, 2 to 3 feet tall, the slender glossy whitish **stems** are very smooth to the touch and are hollow. **Leaves** not numerous, 1 to 2 inches long, ½ inch wide at the base, tapered to a very slender tip, the margins are deeply pinnately lobed. **Flowers** ¼ inch across, the tiny sepals and petals borne at the tip of the long slender hispid hypanthium. Petals ⅛ inch long, pale yellow.

FLOWERING Spring to summer.
ELEVATION Up to 7,000 feet.
WHERE FOUND Common on sandy soils on plains and along washes.

CLAIRE

LOASACEAE Blazingstar Family

LONG-LOBE BLAZINGSTAR
Mentzelia longiloba J. Darl.

SYNONYM *Mentzelia multiflora* (Nutt.) A. Gray

DESCRIPTION Shrubby perennial, branched throughout, the branches nearly white and harsh to the touch. **Leaves** 1 to 2 inches long, pinnately lobed, thickened and quite brittle. The surfaces are covered with a close pubescence which causes them to stick to anything that touches them. **Flowers** 1 to 1½ inches across, the ten bright yellow petals and numerous yellow stamens very attractive. The petals are all cupped upward. Usually there are not many flowers in bloom at one time. Buds are light salmon color.

FLOWERING Summer.

ELEVATION Up to 7,500 feet.

WHERE FOUND Open sunny areas in dry, sandy or rocky, well-drained soil, gravel bars, roadsides, creosotebush scrub.

RACHEL STRINGHAM

MALPIGHIACEAE Barbados-Cherry Family

SLENDER JANUSIA

Janusia gracilis A. Gray

SYNONYM *Cottsia gracilis* (Gray) W.R. Anderson & C. Davis

DESCRIPTION Almost a shrub in appearance, to 2 feet high, the **stems** twining around each other and nearby shrubs. **Leaves** very dark green, slender and about ¾ to 1 inches long. **Flowers** very dainty, ½ inch across, bright yellow, turning red-brown in age. The petals are very slender and broader at the ends than at the bases. **Fruits** bright orange-red, and in two sections which are spreading.

FLOWERING Spring to fall.

ELEVATION Up to 5,000 feet.

WHERE FOUND Common on dry, rocky slopes.

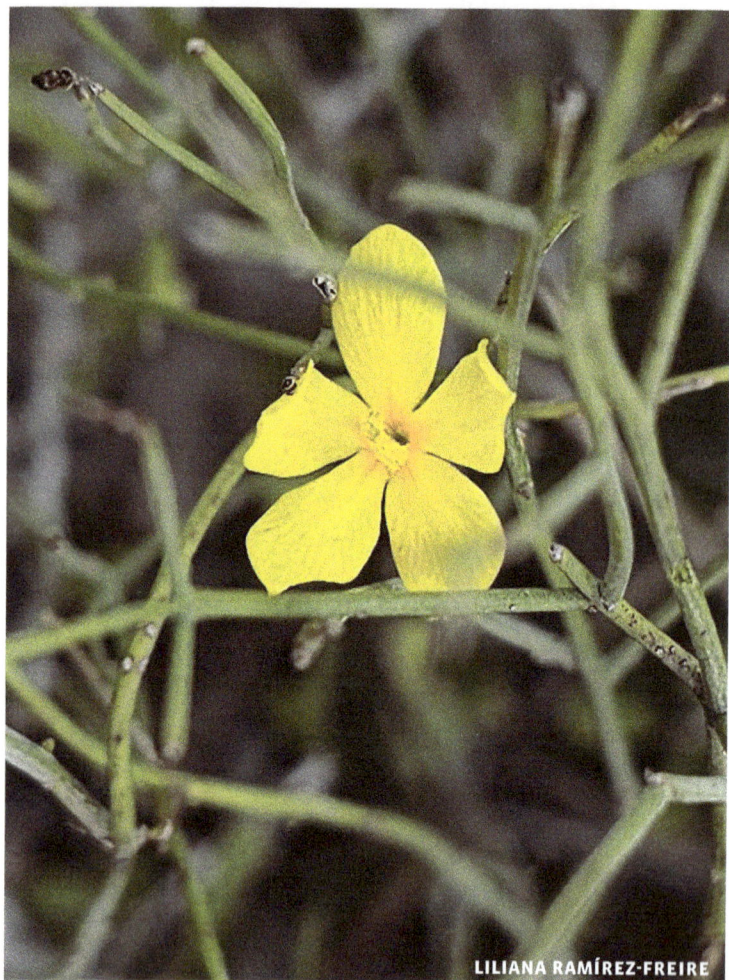

LILIANA RAMÍREZ-FREIRE

OLEACEAE Olive Family

ROUGH MENODORA

Menodora scabra Engelm. ex A. Gray

DESCRIPTION Branched throughout, 12 to 15 inches high, **stems** slender, light green, scabrous-puberulent. **Leaves** alternate, ¾ to 1 inches long, narrow, light green, scabrous-puberulent. **Flowers** few, ¾ to 1 inches across, very showy, bright yellow.

FLOWERING Late spring to summer.

ELEVATION Up to 7,000 feet.

WHERE FOUND Occasional on dry mesas, canyons, slopes, rocky hillsides.

NOTE These flowers are very similar to those of the common yellow jessamine in cultivation.

MATT BERGER

ONAGRACEAE Evening-Primrose Family

CALIFORNIA-PRIMROSE

Eulobus californicus Nutt.

DESCRIPTION Tall, one slender **stem** from the base, this branched above, the branches smooth and green or gray-green. **Leaves** several inches long, very slender, the margins coarsely toothed. Only a few flowers open at one time, these small, yellow. The four green sepals are directed back from the open flower. **Pods** 2 to 3 inches long, very slender.

FLOWERING Spring.

ELEVATION Below 4,500 feet.

WHERE FOUND Common in cool moist places.

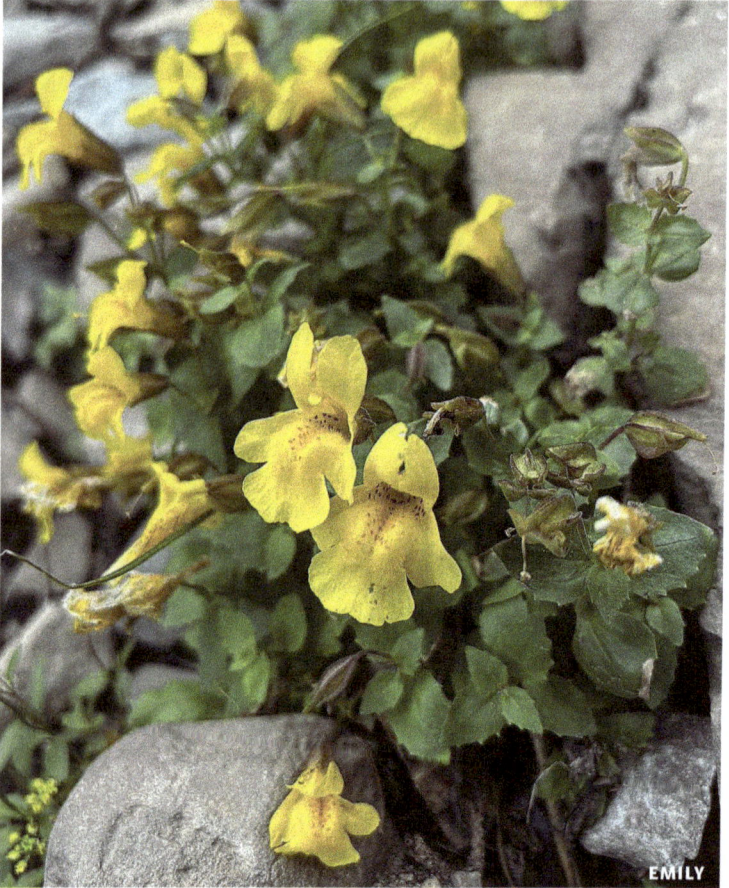

EMILY

PHRYMACEAE Lopseed Family

COMMON YELLOW MONKEY-FLOWER
Erythranthe guttata (DC.) G.L. Nesom

SYNONYM *Mimulus langsdorffii* Don.

DESCRIPTION **Stems** stout, somewhat hollow, 6 to 12 inches high. **Leaves** 2 to 3½ inches long, the blades 1 to 1¾ inches across, ovate, coarsely toothed, yellow-green to green or brown. **Flowers** ¾ to 1 inches long, bright yellow, very much like those of the snapdragon; calyx of five united sepals.

FLOWERING Spring.

ELEVATION Up to 9,000 feet.

WHERE FOUND Common, growing in very moist places, preferably at the edges of pools of water; mostly terrestrial or may be floating in mats.

NOTE A conspicuous plant with showy yellow flowers, sometimes used for salad and greens.

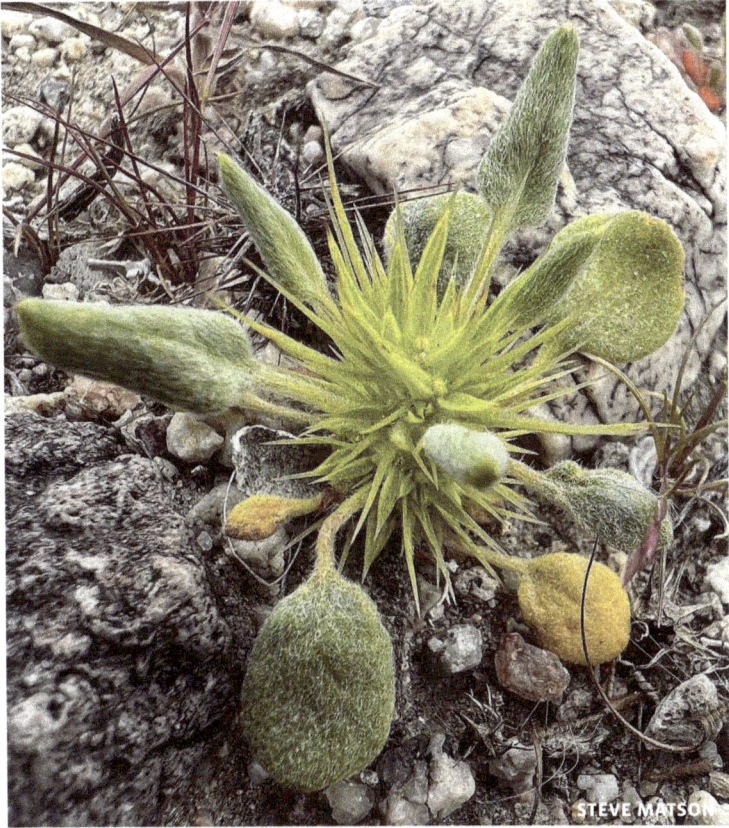

STEVE MATSON

POLYGONACEAE Buckwheat Family

DEVIL'S SPINEFLOWER
Chorizanthe rigida (Torr.) Torr. & A. Gray

DESCRIPTION Seldom more than 2 to 3 inches high, one **stem** from the base or this with one or rarely two branches. Spines or bracts many and closely crowded, linear-lanceolate and tipped with stout spines. **Leaves** few, oblanceolate, pale green. **Flowers** small, yellow.

FLOWERING Spring.

ELEVATION Up to 2,500 feet.

WHERE FOUND Very common in dry sandy or rocky flats and slopes, desert scrub.

NOTE The blackened plants (*right*) are persistent and very conspicuous long after they have died.

AMBER M. KING

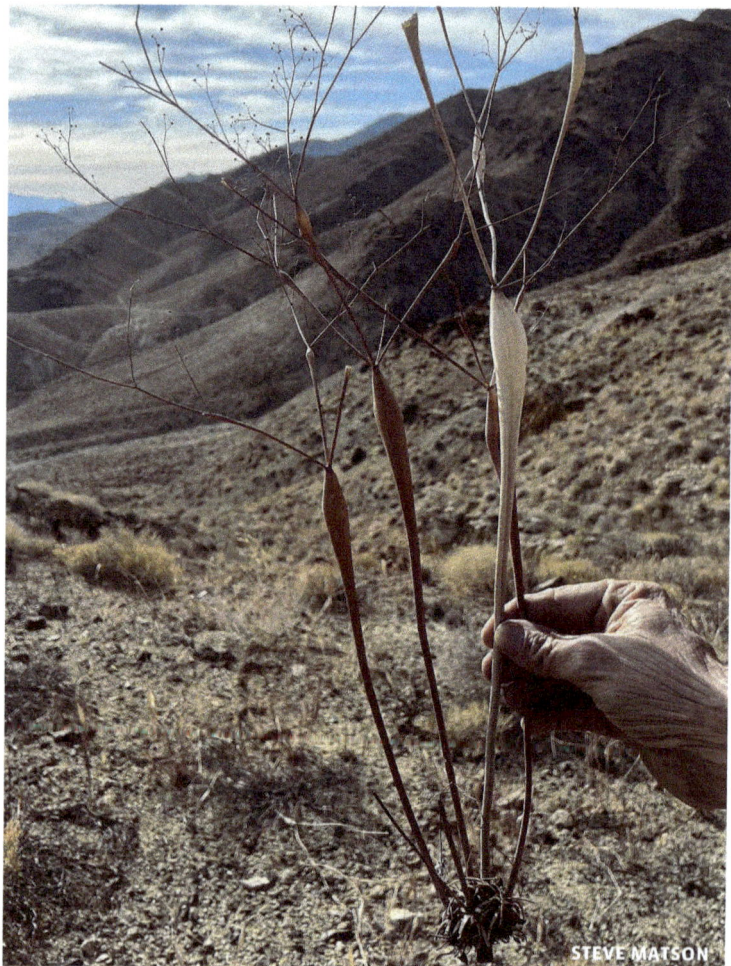

STEVE MATSON

POLYGONACEAE Buckwheat Family

DESERT TRUMPET
Eriogonum inflatum Torr.

DESCRIPTION Annual, 2 to 3 feet high, one or two stems from a basal rosette of dark green, oblong, somewhat wrinkled leaves 1 to 2 inches long. The main **stems** are inflated to as much as ½ inch just below where they branch. The branches are also inflated. **Flowers** small, yellow, rather inconspicuous and not numerous.

FLOWERING Spring to summer.

ELEVATION Up to about 3,500 feet or more.

WHERE FOUND Common on rocky foothills, lower slopes of desert mountains, dry sand or gravel.

NOTE The inflated stem is distinctive.

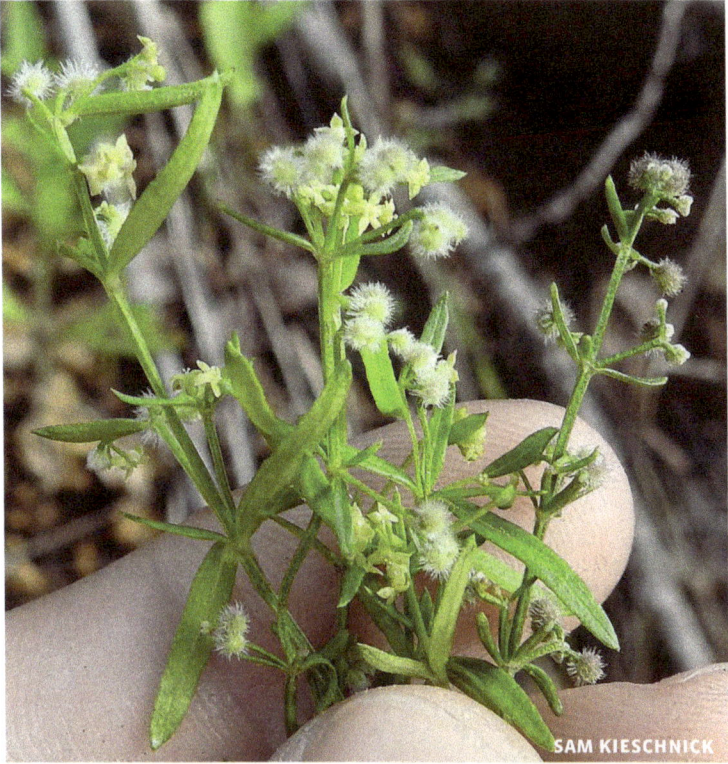

SAM KIESCHNICK

RUBIACEAE Madder Family

FENDLER'S BEDSTRAW
Galium fendleri A. Gray

DESCRIPTION Perennial, 18 to 24 inches tall, intricately branched throughout. **Stems** four-angled, the internodes very short, the nodes breaking easily. **Leaves** ¼ to ⅜ inches long, entire, narrowly elliptical, dark green. Leaves and stems scabrous with short rough hairs. **Flowers** numerous, axillary, ⅛ inches across, yellow. **Fruit** ¹⁄₁₆ inch long, entirely covered with short white hairs.

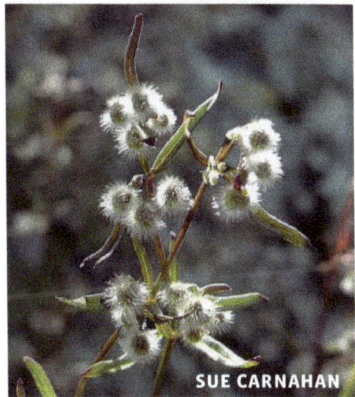

SUE CARNAHAN

FLOWERING Spring.

ELEVATION 8,000 to 9,500 feet.

WHERE FOUND Common among rocks in conifer forests.

TYLER BAILEY

SOLANACEAE Potato Family

TREE-TOBACCO
Nicotiana glauca Graham

DESCRIPTION Tree, 6 to 12 feet tall, branched throughout, the branches heavy and dark green. **Leaves** 6 to 15 inches long, oblong, heavy and leathery, dark green and covered with a powdery white bloom. **Flowers** many, in large clusters; flowers 1½ to 2 inches long, tubular, yellow-green.

FLOWERING Spring.

ELEVATION Up to 3,000 feet.

WHERE FOUND Common, usually in rather damp places as along streams, ditches, washes; open disturbed flats and slopes.

MELANIE DAVIS

SOLANACEAE Potato Family

DESERT TOBACCO
Nicotiana obtusifolia M. Martens & Galeotti

SYNONYM *Nicotiana trigonophylla* Dunal.

DESCRIPTION Coarse, 1 to 2 feet tall, somewhat branched; **stems** heavy. **Leaves** 2 to 5 inches long, ¾ to 1½ inches wide, broadly lanceolate to oblanceolate, yellow green. Stems and leaves short rough-hairy. **Inflorescence** a raceme, flowers ¾ inch long, corolla funnelform, slightly constricted at the tip under the five very short somewhat spreading corolla lobes. **Flowers** yellow-green. **Fruit** a capsule.

FLOWERING Spring.

ELEVATION Up to 6,000 feet, usually lower.

WHERE FOUND Common along sandy, gravelly or rocky washes, slopes.

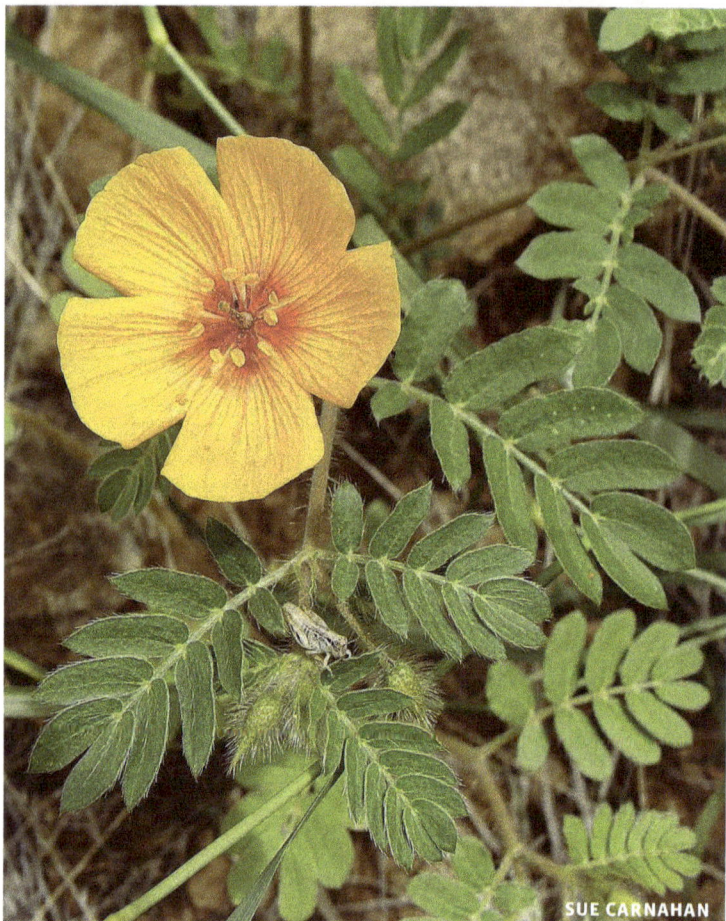

SUE CARNAHAN

ZYGOPHYLLACEAE Creosotebush Family

ARIZONA POPPY, ORANGE CALTROP

Kallstroemia grandiflora Torr. ex A. Gray

DESCRIPTION Procumbent or partly erect , similar in appearance to the **puncture-vine** (*Tribulus terrestris*). **Leaves** 1½ to 2 inches long, dark green, coarse hirsute. **Flowers** large, 1 inch or more across, bright orange. **Fruits** roughened and somewhat tuberculate but not spiny.

FLOWERING Spring to fall.

ELEVATION Up to 5,000 feet.

WHERE FOUND Common along roadsides and on and mesas.

NOTE This plant, often locally known as Arizona-poppy, is one of the most attractive summer annuals found in the southern part of the State. The large flowers, rich orange in color, superficially resemble those of the California poppy (*Eschscholzia*).

ERIC IN SF

ZYGOPHYLLACEAE
Creosotebush Family

CREOSOTEBUSH

Larrea tridentata (DC.) Coville

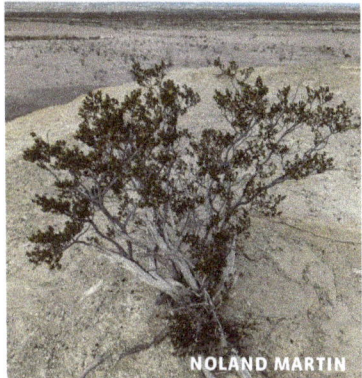

NOLAND MARTIN

DESCRIPTION Strongly scented resinous shrub; many spreading branches from the base. **Leaves** two-foliolate, dark glossy green. **Flowers** many, bright yellow, the petals twisted somewhat like the blades of a propeller. **Fruit** globose, densely covered with cream-white hairs.

FLOWERING Spring to summer.

ELEVATION Up to 5,000 feet.

WHERE FOUND Very common in desert scrub.

NOTE A very important part of the perennial desert flora in southern and western Arizona. Creosotebush covers thousands of square miles, often in nearly pure stands, and usually with remarkably little variation in size. The Pima Indians used the leaves in decoction as an emetic and to poultice sores. Small quantities of "lac" are found on the branches as a resinous incrustation. This was used for fixing arrow points, mending pottery, etc. creosotebush has a strong characteristic odor, especially noticeable when the leaves are wet. The plant ordinarily is not touched by livestock although it is reported that sheep, especially pregnant ewes, have been killed by eating it. This plant is reported to cause dermatitis in some persons who are allergic to it.

SKJOLD SØNDERGAARD

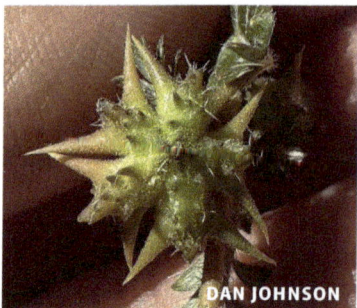

DAN JOHNSON

ZYGOPHYLLACEAE
Creosotebush Family

PUNCTURE-VINE
Tribulus terrestris L.

DESCRIPTION Procumbent annual, many **stems** from the base, spreading 18 to 24 inches from the center of the plant, coarsely hirsute. **Leaves** 1 to 1½ inches long, pinnately divided, leaflets six pairs, ¼ inch long, hairy. **Flowers** yellow, ⅜ inch across. **Fruit** of five carpels, each with two wide-spreading spines.

FLOWERING Spring to fall.

ELEVATION Up to 6,000 feet.

WHERE FOUND Introduced and often abundant on roadsides, fields, vacant lots, railways and disturbed areas; a weed in cultivated places.

NOTE A troublesome annual because of the numerous spiny fruits each plant produces. Although the plants produce seeds in great quantities, and the fruits are readily disseminated by means of furred animals and automobile tires, it is easily controlled by cultivation. The spines of the fruit are not capable of puncturing automobile tires but do penetrate bicycle tires. A single carpel somewhat resembles the head of a bull, leading to another common name of "bullhead." Hay containing the fruits may cause troublesome sores in the mouths of livestock.

RACHEL STRINGHAM

APIACEAE Carrot Family

HOARY BOWLESIA

Bowlesia incana Ruiz & Pav.

DESCRIPTION A delicate trailing annual found in shady and moist places especially along stream beds and under large plants. The **stems** are slender, as much as 2 feet long, the pale green leaves are opposite and quite far apart. The **leaves** are about 1 inches wide, rounded and with five to seven deep lobes on the margins. The stems and leaves have fine stellate hairs on the surface. **Flowers** white, very small and inconspicuous, but the ovate, two-lobed **fruit** is nearly ⅛ inch long, light green when young and light brown when older.

FLOWERING Spring.

ELEVATION Up to 3,500 feet.

WHERE FOUND Common, usually among shrubs, in the shade of trees, rocks.

DUTZA K.

SUE CARNAHAN

APIACEAE Carrot Family

AMERICAN WILD CARROT
Daucus pusillus Michx.

DESCRIPTION Slender erect annual, 3 to 5 inches tall, more or less branched throughout. **Leaves** 1½ to 2½ inches long, feathery, dark green, the leaf divisions very small. **Flowers** tiny, white, in a more or less irregular cluster at the tip of the plant. **Fruit** 3/32 inch long, with small barbed prickles arranged in rows on the sides.

FLOWERING Spring.

ELEVATION Up to 4,000 feet.

WHERE FOUND Occasional in rocky or sandy places.

NOTE It is reported that the Navajo Indians ate the roots, both raw and cooked.

ALISON NORTHUP

APIACEAE Carrot Family

BRISTLY-FRUIT SCALESEED

Spermolepis echinata (Nutt. ex DC.) A.Heller

DESCRIPTION Slender erect, 3 to 5 inches tall, sparingly branched, the leaves 1½ inch long, very deeply dissected into long narrow lobes similar to those of the carrot (*Daucus*). The small white **flowers** are borne in a flat cluster at the tip of the stem and quite easily seen. The individual flowers are only ¹/₁₆ inches across (or smaller). The **fruit** is very small, and covered with minute hooked prickles. The plant when crushed has a faint carrot odor.

FLOWERING Spring.

ELEVATION Up to 5,000 feet.

WHERE FOUND Occasional on rocky slopes, sandy flats.

KLICKLO

ASTERACEAE Aster Family

DESERT BROOM, ROSINBUSH

Baccharis sarothroides A. Gray

DESCRIPTION Shrub, 6 to 8 feet high, many branches from the base and densely branched throughout, giving the plant a broom-like appearance. Branches yellow-green and angled. **Leaves** alternate, ¼ to 1 inches long, entire. **Flower heads** very many, ⅛ inch long, involucral bracts overlapping, dull green; **flowers** cream-colored, pappus abundant, elongating in fruit.

FLOWERING Summer.

ELEVATION Up to 5,000 feet.

WHERE FOUND Common in sandy and gravelly washes, railroads, roadsides, flooded-areas, disturbed places; sometimes in saline soil.

MATT BERGER

ASTERACEAE Aster Family

DESERT PINCHUSHION, MORNING BRIDES
Chaenactis stevioides Hook. & Arn.

DESCRIPTION Annual, 8 to 15 inches high, few branches from near the base.
Stems slender, green to purplish, puberulent. **Leaves** 1 to 2 inches long,
pinnately divided, the divisions very slender, almost thread-like, ¼ to
½ inches long. **Flower heads** numerous, ½ inch long, involucre of numerous slender green bracts; **flowers** disk only, about ⅛ inch across
and creamy white.

FLOWERING Spring.

ELEVATION To about 6,500 feet.

WHERE FOUND Common on sandy, gravelly and rocky slopes, washes and
mesas.

MATT BERGER

ASTERACEAE Aster Family

WHITE EASTERBONNETS
Eriophyllum lanosum (A. Gray) A. Gray

SYNONYM *Antheropeas lanosum* (A. Gray) Rydb.

DESCRIPTION Annual, 2 inches to 3 inches tall, few branches from the base. Leaves ¼ inch long, oblanceolate, gray-green. Leaves and stems covered with a dense white down. Flower heads few, ½ inch across, rays eight to ten, ³/₁₆ inch long, toothed at the tips, white; disk ¼ inch across, yellow.

FLOWERING Spring.

ELEVATION To about 3,000 feet.

WHERE FOUND Common on dry, gravelly mesas and slopes.

ADAM J. SEARCY

SUE CARNAHAN

ASTERACEAE Aster Family

CROWFOOT ROCK DAISY

Galinsogeopsis coronopifolia (A. Gray) Lichter-Marck

SYNONYM *Perityle coronopifolia* A. Gray

DESCRIPTION Low slender perennial, 12 to 15 inches high, many branches from the base and throughout. **Leaves** various, 1 to 1¾ inches long, the blades pinnately or palmately lobed, the lobes sharp. Lower leaves opposite. **Flower heads** many, ½ inch across, rays about ten, ³/16 inch long, white; disk ¼ inch across, yellow.

FLOWERING Spring.

ELEVATION 4,000 to 7,500 feet.

WHERE FOUND Common among rocks and on cliffs, often on limestone.

MATT BERGER

ASTERACEAE Aster Family

DESERT CHICORY, SNAKE WEED

Rafinesquia neomexicana A. Gray

SYNONYM *Nemoseris neomexicana* (A. Gray) Greene

DESCRIPTION Rather low, 8 to 18 inches high, branched from or near the base and bearing numerous large, flat, white **flowers** about 1½ inches across. The **stems** are filled with a milky juice.

FLOWERING Spring to fall.

ELEVATION Up to about 3,500 feet.

WHERE FOUND Very common on the lower slopes of the mountains.

NOTE Sometimes called "snake weed" because of the similarity of the bracts at the base of the flower head to the scales of a snake.

MATT BERGER

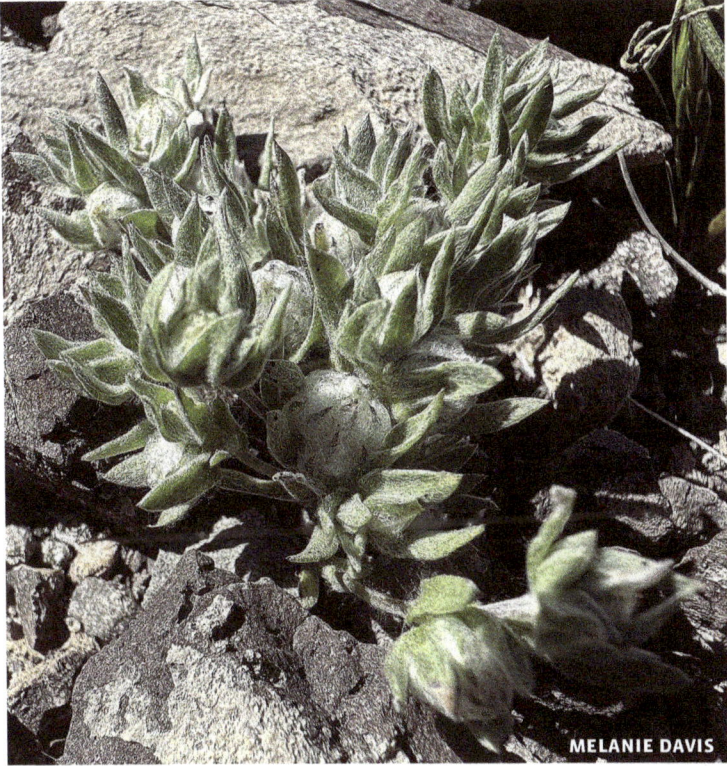

MELANIE DAVIS

ASTERACEAE Aster Family

WOOLLY-HEAD NESTSTRAW
Stylocline micropoides A. Gray

DESCRIPTION Annual, 3 to 4 inches high, **stems** slender, brittle, brown. **Leaves** ⅜ inch long, thread-like. Leaves and stems covered with a close white down. **Flower heads** very numerous, the flowers themselves very inconspicuous but the involucral bracts very showy. Bracts ⅛ inch long, each bearing a ball of "cotton." When all together the flower head has the appearance of a cotton boll ⅜ to ½ inches across.

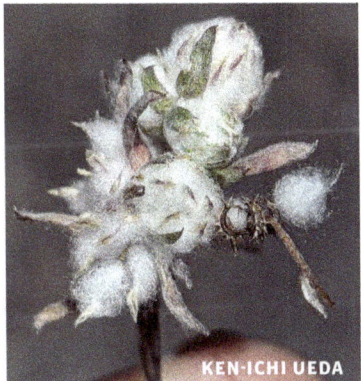

KEN-ICHI UEDA

FLOWERING Spring.

ELEVATION To 4,000 feet.

WHERE FOUND Common on sandy or gravelly soil.

ALAN PRATHER

BORAGINACEAE Borage Family

BROAD-FRUIT COMBSEED

Pectocarya platycarpa (Munz & I.M. Johnst.) Munz & I.M. Johnst.

DESCRIPTION Never more than 6 inches high, with many branches from the base and with rather long narrow green leaves. The branches are somewhat spreading. The entire plant is covered with short rough hairs. The small white **flowers**, about $^1/16$ inch across, are quite inconspicuous but the **fruits**, from which the plant gets its common name, split open in four sections; the sections curl backwards slightly, are light cream-yellow and have short teeth along the sides.

FLOWERING Spring.

ELEVATION Up to 5,000 feet (mostly lower).

WHERE FOUND Very common on rocky slopes and in open places.

BORAGINACEAE Borage Family

ARIZONA POPCORN-FLOWER

Plagiobothrys arizonicus (A. Gray) Greene ex A. Gray

DESCRIPTION Rather coarse, many stems from the base, 4 to 6 inches high. **Leaves** 1 to 3 inches long, ¼ inch wide, medium green, the veins filled with a red-purple juice; densely hispid on the upper surface, much less so below. The lower stems and root are also filled with this juice. **Inflorescence** a short closely coiled helicoid cyme. **Flowers** ⅛ inch across, white.

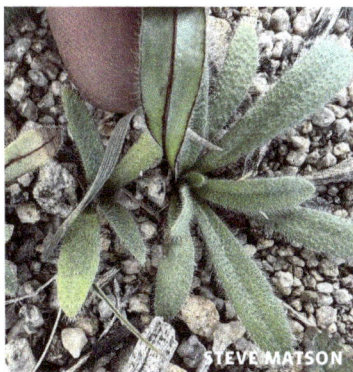

FLOWERING Spring.

ELEVATION Up to 4,000 feet, more abundant at lower elevations.

WHERE FOUND Common between bushes and rocks in creosotebush scrub and oak woodlands.

MATT BERGER

BRASSICACEAE Mustard Family

TOURISTPLANT
Dimorphocarpa wislizeni (Engelm.) Rollins

SYNONYM *Dithyrea wislizeni* Engelm.

DESCRIPTION Rather coarse, 12 to 24 inches tall, many **stems** from the base, these erect or nearly so and covered with a fine white stellate pubescence. **Leaves** 2 to 3 inches long, linear-lanceolate, irregularly toothed, pale green and with the same stellate pubescence as the stems. **Flowers** numerous on the elongated raceme. Flowers ⅜ to ½ inches across, white with pale yellow stamens. The **pods** are flattened, ¼ inch long and ½ inch or more in width. The two halves are rounded in such a way as to resemble a small pair of spectacles. There is one seed in each half of the pod.

FLOWERING Spring.

ELEVATION Up to 6,000 feet.

WHERE FOUND Very common along sandy river beds.

SAM KIESCHNICK

BRASSICACEAE Mustard Family

HAIRY-POD PEPPERWORT
Lepidium lasiocarpum Nutt.

DESCRIPTION Much branched throughout, 4 to 8 inches tall, the stems stout. **Leaves** ¾ to 1¼ inches long, oblong to oblanceolate. **Flowers** minute, inconspicuous, white. **Pods** very numerous, flattened, ⅛ inches across, each with a shallow indentation at the tip. The entire plant is finely hirsute.

FLOWERING Early spring.

ELEVATION Up to 4,000 feet.

WHERE FOUND Very common on sandy soil.

CHLOE AND TREVOR VAN LOON

BRASSICACEAE Mustard Family

MOUNTAIN FRINGEPOD

Thysanocarpus laciniatus Nutt.

DESCRIPTION Slender, 12 to 18 inches tall. **Leaves** 1½ to 2 inches long, dark green, very slender, rarely somewhat toothed at the bases. **Flowers** tiny, white or cream color. The **pods** are flat, ¼ inch long, the margins perforated and with the appearance of having been hem-stitched; these pods hang down on the slender pedicels and are very attractive.

FLOWERING Spring.

ELEVATION 3,500 to 5,000 feet.

WHERE FOUND Occasional, only in moist places.

NOTE Mountain fringepod is one of the most dainty of the mustards.

MEL LETTERMAN

BURSERACEAE Frankincense Family

ELEPHANT-TREE
Bursera microphylla A. Gray

DESCRIPTION Tree, branched throughout, the branches very limber, the older ones with a dark red-brown bark that peels off readily. **Leaves** very dark green and when crushed giving off a strong spicy odor which resembles that of the pepper trees found in Phoenix. **Flowers** tiny, the sepals inconspicuous, the waxy white petals only 3/32 inch long; the flowers have a faint sweet odor. **Fruit** three-angled, ¼ inch long, dark brown.

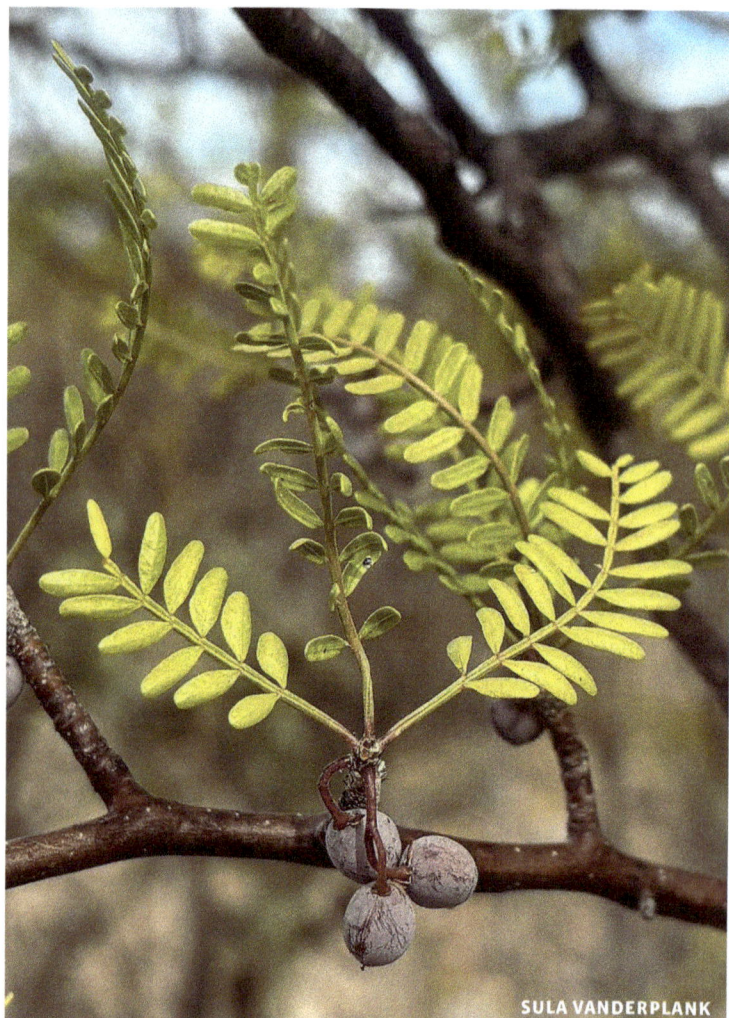

SULA VANDERPLANK

FLOWERING Summer.

ELEVATION Up to 2,500 feet.

WHERE FOUND Common locally on arid rocky slopes and limestone soils.

NOTE This tree is rather common in the South Mountains and at a distance is sometimes mistaken for mesquite. The trees reach a height (in Arizona) of 20 feet and a trunk diameter of 1 foot. The crooked branches taper rapidly, resembling the trunk of an elephant. The plant cannot withstand much cold. The bark contains tannin and was gathered in Sonora for export. In that region the gum was used for treating venereal diseases.

SUSAN MURPHY

CACTACEAE Cactus Family

SAGUARO

Carnegiea gigantea (Engelm.) Britton & Rose

DESCRIPTION **Stem** simple and upright to 50 feet in height, with one or more lateral branches; ribs twelve to twenty-four, ¾ to 1½ inches high, areoles 1 inch apart or nearly contiguous on the upper part of the plant, densely brown felted; **spines** of two kinds, those at the top or the flowering stems needle-shaped, yellowish brown, those of the sterile plants and on the lower parts of the flowering plants more or less awl-shaped, the central ones stouter than the radials, often 2½ inches long. **Flowers** 4 to 5 inches long, sometimes nearly as broad as long when fully expanded, waxy white. **Fruit** a red or purple berry, 3 inches long, fleshy, edible.

FLOWERING Spring.

ELEVATION Up to 4,500 feet (usually under 3,500 feet).

WHERE FOUND Common on rocky hills and plains in well-drained soils.

NOTE Arizona's State flower, and the largest cactus in the state, occasionally attaining a height of more than 50 feet and developing as many as 50 "arms." Large individuals are believed to be from 150 to 200 years old, or more. The flowers are nocturnal, opening between 9 and 12 o'clock at night. They open slowly, full expansion requiring about 2 or 3 hours;

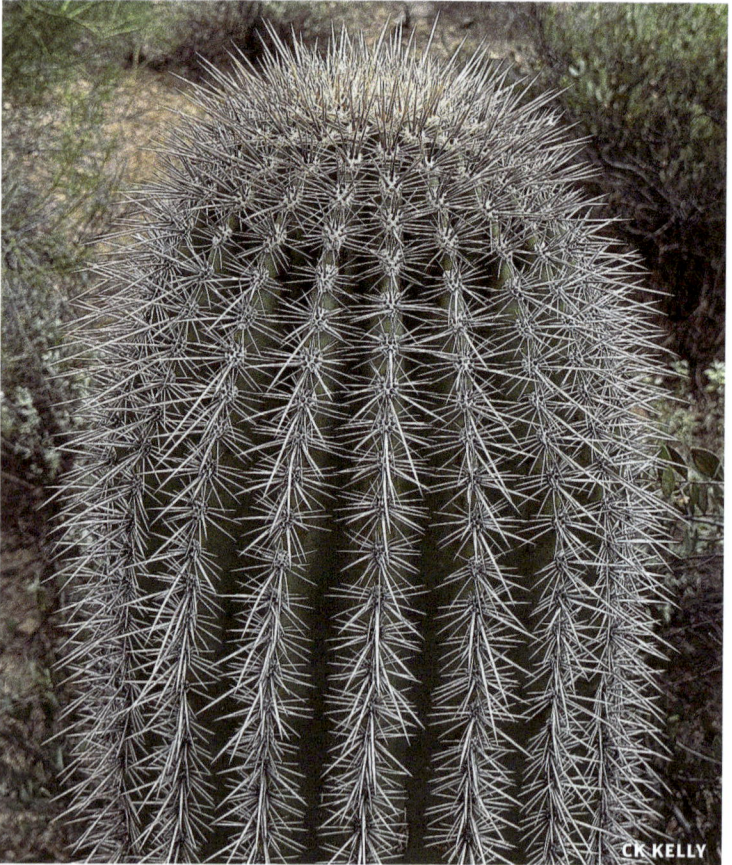

CK KELLY

and persist in full flowering stage until late the following afternoon when they begin to close and wither. The large and beautiful white flowers are not fragrant but have an odor like that of ripe melon.

The saguaro has contributed substantially toward the subsistence of the Pima and Papago Indians, furnishing materials for food and shelter. Their great capacity for storing water, combined with slow rate of growth, enables the plant to fruit annually more or less irrespective of drought. The fruit, or "pitahaya" of the early Spaniards, matures in June and July. The watermelon-red pulp is eaten fresh or stored, and in the form of syrup and preserves. At the annual harvest an intoxicating beverage is prepared by allowing the juice to ferment. The Papagos make a sort of butter from the seeds. The white-wing dove, a favorite game bird of Arizona hunters, feeds largely on the seeds of the saguaro during the fruiting season.

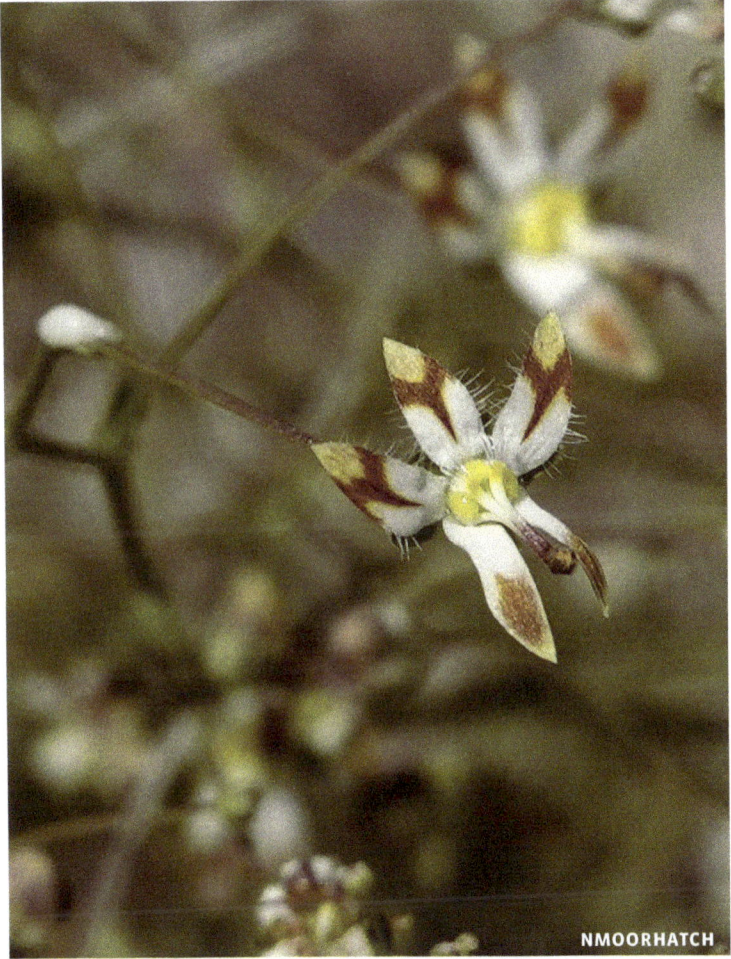

NMOORHATCH

CAMPANULACEAE Bellflower Family

DESERT THREADPLANT
Nemacladus rubescens Greene

DESCRIPTION Low growing annual, 2 to 6 inches tall, branched throughout, the branches slender, green to brown. **Leaves** ¼ inch long, narrowly linear, green. **Flowers** numerous, ³/16 inch across, white, the two upper lobes erect, the three lower ones bent slightly downward, each bearing a small purple dot near the base. Anthers deep purple.

FLOWERING Spring.

ELEVATION Up to about 2,000 feet.

WHERE FOUND Occasional on dry, gravelly or rocky soils.

NOTE A well-marked species, easily distinguished from all others by the silvery gray stems and the smooth, yellowish-green, nearly entire leaves.

LAURA GAUDETTE

CROSSOSOMATACEAE Rockflower Family

RAGGED ROCKFLOWER

Crossosoma bigelovii S. Watson

DESCRIPTION Shrub 2 to 3 feet high, spreading, the stems slender, with a dull gray rough bark which is very bitter if chewed. **Leaves** alternate, elliptical, and not exceeding ½ inch in length, medium green, glabrous, very bitter. **Flowers** cream-white, ¾ inch across, stamens numerous, yellow.

FLOWERING Spring.

ELEVATION 1,500 to 4,000 feet.

WHERE FOUND Common on dry rocky slopes, cliffs and canyon walls.

NOTE A straggling shrub, with very astringent bark and white flowers, these sometimes tinged with purple. Worthy of cultivation on account of the appealing fragrance of the flowers.

MATT BERGER

EUPHORBIACEAE Spurge Family

YUMA SILVERBUSH

Argythamnia serrata (Torr.) Müll.-Arg.

SYNONYM *Ditaxis neomexicana* (Müll.-Arg.) Heller.

DESCRIPTION Shrubby, 1 to 2 feet high, with a few branches. Lower stems with a light gray bark, upper branches and stems bright green and sparsely covered with an appressed silky pubescence. **Leaves** alternate, with very short petioles; leaf blades ¾ to 1 inches long, lanceolate, pale green, both sides covered with an appressed white silky pubescence. **Flowers** numerous, clustered at the ends of the small branches; flowers ³/16 inches across, petals 3/32 inch long, obovate, pale cream-color; anthers yellow. **Fruit** three-parted, covered with a silky pubescence.

FLOWERING Spring to fall.

ELEVATION Up to 3,500 feet.

WHERE FOUND Common on sandy and rocky slopes, along washes, in creosotebush scrub.

SUE CARNAHAN

FABACEAE Pea Family

VELVET MESQUITE

Prosopis velutina Woot.

SYNONYM *Neltuma velutina* (Woot.) Britt & Rose

DESCRIPTION Tree 6 to 10 feet high, with a short heavy rough trunk or several heavy branches from the base. Limbs rough and dark green-brown. **Leaves** 1½ to 2½ inches long, leaflets about twelve pairs, ¼ to ½ inches long, oblong, green or yellow-green, hairy. At the base of each leaf cluster are two heavy straight spines, these may be as long as 1½ inch. **Inflorescence** a dense spike 1½ to 3 inches long and ½ inch across. **Flowers** small, cream-colored. **Pods** 4 to 6 inches long, slender, glabrous, yellow when ripe.

FLOWERING Spring.

ELEVATION Mostly below 4,500 feet.

WHERE FOUND Common in lower desert plant communities that may include palo-verde (*Parkinsonia*), bursage (*Ambrosia*), various species of cactus and shrublands that include pinyon-juniper, pine and oak species. Soils typically sandy or rocky. When growing in riparian areas, velvet mesquite forms thick, dense stands ("bosques") that are important wildlife sanctuaries.

SIMILAR SPECIES **Honey mesquite** (*Prosopis glandulosa* Torr.) is similar, but with glabrous or glabrate foliage and leaflets commonly more than 5/8 inch long.

NOTE It is reported that the roots sometimes penetrate to a depth of 60 feet. The foliage, and particularly the pods, are eaten by livestock. The sapwood is yellow, the heavy reddish brown heartwood hard and slow burning. With the exception of desert-ironwood (*Olneya tesota*), mesquite is the best firewood obtainable in this region. Trees cut to the ground resprout. The wood is used for fence posts and the heartwood is said to take a fine polish. Mesquite increases rapidly on overgrazed grassland in southeastern Arizona and is a serious range pest under such circumstances.

This plant has been a mainstay of existence to the native peoples of the Southwest. When cultivated crops failed, the Indians subsisted mainly upon mesquite beans. Pinole, a meal made from the long sweet pods, prepared in the form of cakes and in other ways, was a staple food with the Pimas and still is eaten by them to some extent. Fermented pinole was a favorite intoxicating drink. The gum which exudes from the bark was used to make candy, to mend pottery, and as a black dye. The inner bark furnished the Indians material for basketry and coarse fabrics, as well as medicine to treat a variety of disorders. Under normal conditions large quantities of excellent honey are obtained from the flowers of mesquite, which is rated by beekeepers as the most valuable honey plant of the State.

CHLOE AND TREVOR VAN LOON

LOASACEAE Blazingstar Family

THURBER'S SANDPAPER-PLANT

Petalonyx thurberi A. Gray

DESCRIPTION Biennial or perennial, shrubby and branched throughout, 2 to 3 feet high. **Stems** white and very rough. **Leaves** lanceolate, margins shallowly toothed, surfaces medium green and very scabrous or sand-paper-like. **Flowers** numerous, ¼ inch across, the petals cream- colored; the five stamens are borne at the base of the flower and apparently come out on the outside of the petals.

FLOWERING Spring, summer.

ELEVATION Up to 5,000 feet but usually much lower.

WHERE FOUND Common in sandy river beds.

DAN WRENCH

MALVACEAE Mallow Family

CHEESEWEED MALLOW

Malva parviflora L.

DESCRIPTION Low growing annual, 18 to 24 inches tall, in dry places only 6 to 8 inches tall. **Leaves** nearly round, the margins very shallowly scalloped. **Flowers** small, white, the tips of the petals lavender. **Fruits** disk-shape, called cheeses by children.

FLOWERING Spring.

ELEVATION Up to 5,000 feet.

WHERE FOUND Common introduced weed of crop fields, gardens, urban areas, roadsides and other disturbed places.

STEVE MATSON

NYCTAGINACEAE Four-O'clock Family

DESERT WISHBONEBUSH

Mirabilis laevis (Benth.) Curran

DESCRIPTION A straggling, weak-stemmed plant about 1½ to 2 feet long, with hairy and sticky stems. Plants are usually found growing up into nearby bushes. **Leaves** ovate to bluntly heart-shaped, about 1 inch long, thick, dull green and sticky. **Flowers** about ½ inch across, white or pink, opening at night and gradually closing in the morning.

FLOWERING Spring.

ELEVATION Up to 3,000 feet.

WHERE FOUND Common on rocky slopes.

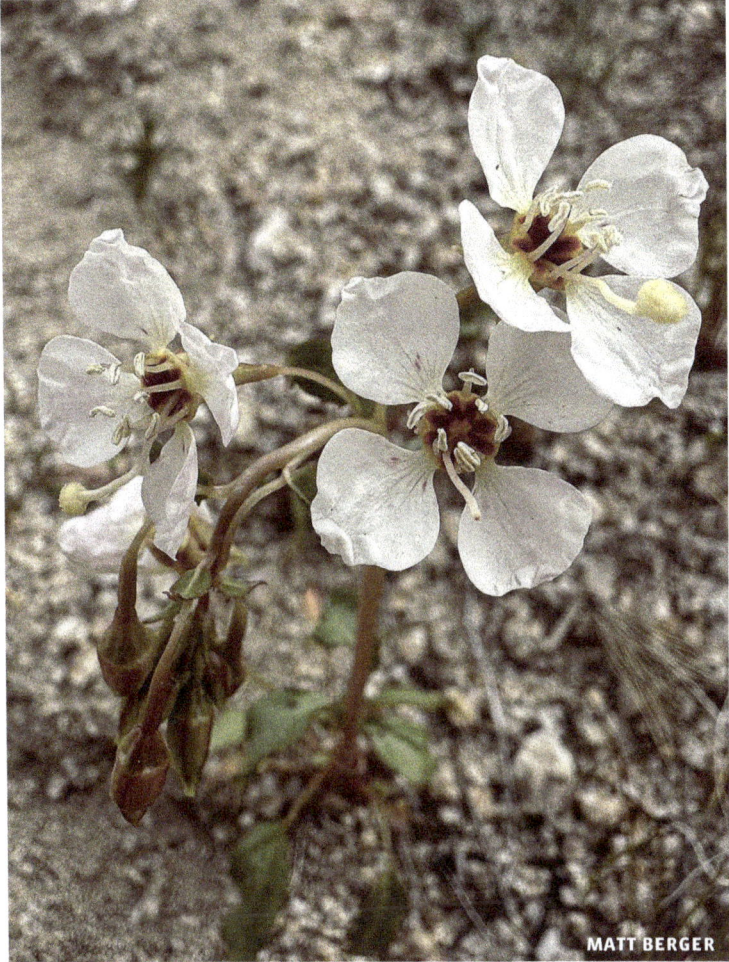

MATT BERGER

ONAGRACEAE Evening-Primrose Family

BROWNEYES

Chylismia claviformis (Torr. & Frém.) A. Heller

DESCRIPTION Low growing, with few branches, these slender, green or dark gray-brown, smooth. **Leaves** 1 to 2 inches long, medium green, at the base deeply pinnately lobed, above merely toothed. **Flowers** numerous, borne in a close raceme. Flowers ½ inch across, cream color, greatly resembling those of the mustards; stamens eight, pale yellow. **Pods** ¾ to 1 inches long, red-brown, when ripe splitting into fourths.

FLOWERING Spring.

ELEVATION Up to 4,500 feet (usually below 3,000 feet).

WHERE FOUND Occasional on rocky slopes, washes and sandy places.

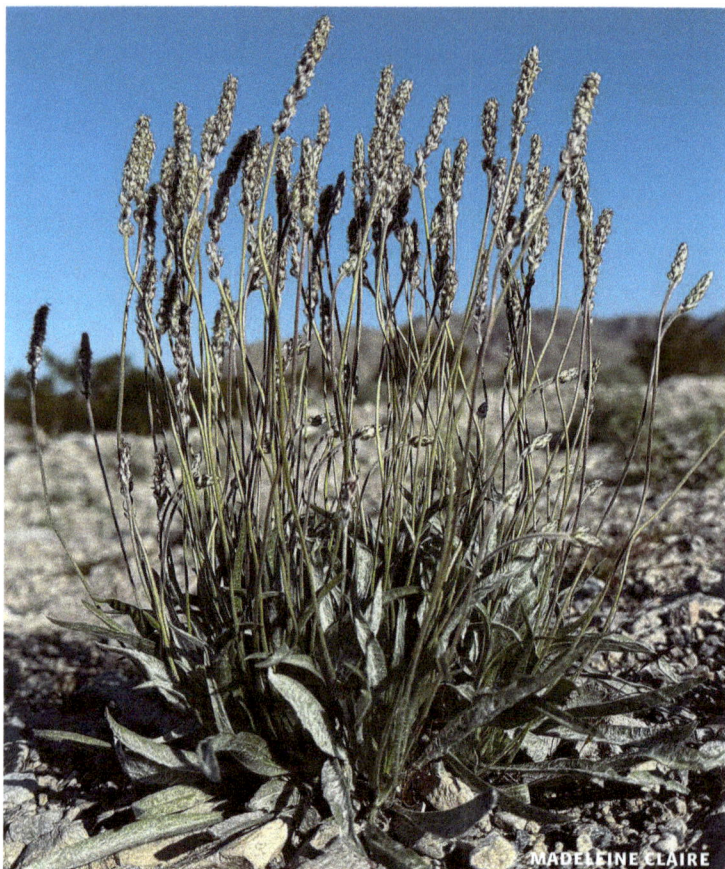

MADELEINE CLAIRE

PLANTAGINACEAE Plantain Family

DESERT PLANTAIN, INDIAN WHEAT
Plantago ovata Forssk.

SYNONYM *Plantago fastigiata* Morris

DESCRIPTION Many slender **stems** from the base, 3 to 8 inches high. **Leaves** numerous, all basal, 1½ to 4 inches long, ⅛ to ¼ inches wide, narrowly linear, yellow-green. Leaves and stems densely covered with soft white hairs. Spikes ½ to 1 inches long, ⅜ inch across, the flowers crowded. **Flowers** ⅛ inches across, the four papery cream-colored lobes bent abruptly back at the base. Calyx lobes linear, finely hairy.

FLOWERING Spring.

ELEVATION Up to 3,000 feet.

WHERE FOUND Very common on dry plains and mesas; soils sandy or gravelly.

NOTE When it first appears in spring, plants cover the ground and have the appearance of a thick grass.

MARK GRONEVELD

POLEMONIACEAE Phlox Family

DESERT-TRUMPETS

Linanthus bigelovii (A. Gray) Greene

DESCRIPTION **Stems** very slender, 4 to 6 inches high, pale to dark brown, glossy, very sparsely short hairy to glabrous. **Leaves** ½ to 1¼ inches long, very slender, dark green. **Flowers** ¾ inch long, corolla bell-shaped, the petals or corolla lobes ⅜ inch long, broad, wide spreading, cream-yellow.

FLOWERING Spring.

ELEVATION Up to 4,000 feet.

WHERE FOUND Common on sandy stream beds, washes, bajadas.

STEVE MATSON

POLYGONACEAE Buckwheat Family

BRITTLE SPINEFLOWER
Chorizanthe brevicornu Torr.

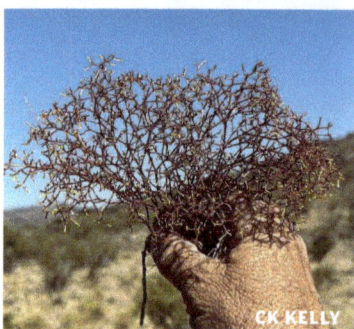

CK KELLY

DESCRIPTION Low annual, 4 to 8 inches high, branched through-out, the **stems** dark green or red, sparsely strigose. **Leaves** basal, ½ to 1 inches long, linear, dark green, sparsely strigose; at each of the upper joints are two reduced leaves. **Flowers** axillary, small, inches across, white. The calyx is $3/16$ inches long, of five or six united sepals, the tips with a short, spreading, spine.

FLOWERING Spring.

ELEVATION Up to about 3,000 feet.

WHERE FOUND Very common on sandy or gravelly soils in various plant associations including grasslands, saltbush, creosotebush, and pinyon-juniper.

NOTE The plant breaks easily at the nodes.

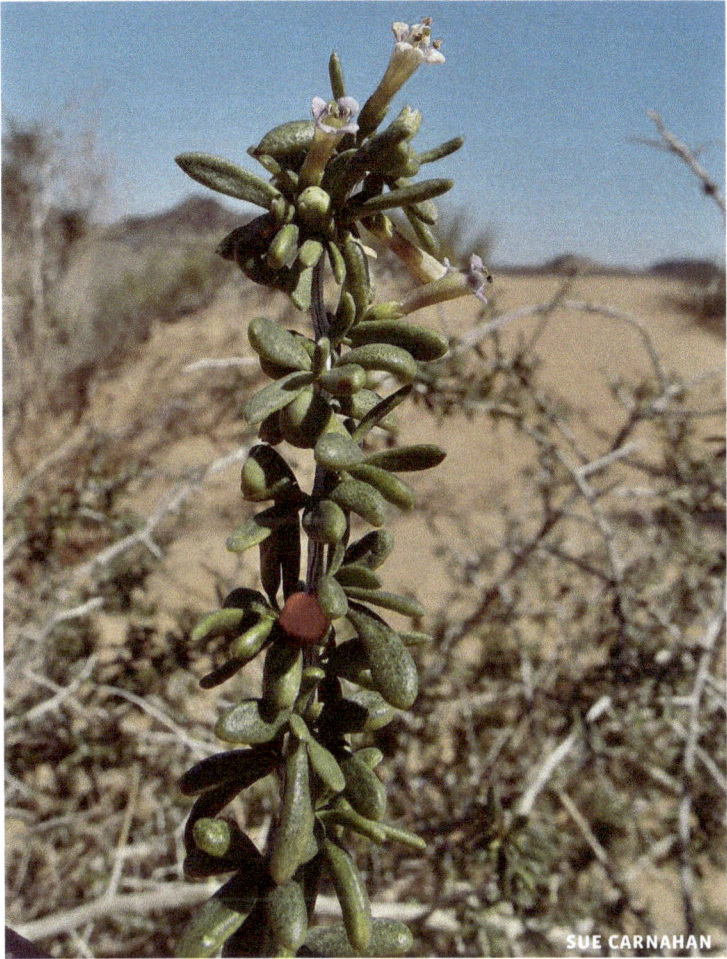

SUE CARNAHAN

SOLANACEAE Potato Family

RED-BERRY DESERT-THORN

Lycium andersonii A. Gray

DESCRIPTION Shrub, 4 to 6 feet high, branched throughout, the smaller branches grayish and at right angles to the larger ones and each tipped with a sharp spine. The **branches** have a soft brown bark that is marked with numerous shallow fissures and is somewhat shreddy. **Leaves** numerous, fascicled, ½ inch long, broadly oblanceolate, rather thick, gray-green, densely puberulent. **Flowers** numerous, to ⅜ inch long, tubular, pale cream color. **Fruit** a berry, ³/16 inch long, globose and nearly black.

FLOWERING Spring.

ELEVATION Up to 5,500 feet.

WHERE FOUND Common along washes.

SUE CARNAHAN

ASPARAGACEAE Asparagus Family

BLUEDICKS

Dipterostemon capitatus (Benth.) Rydb.

SYNONYM *Brodiaea capitata* Benth.

DESCRIPTION The green **leaves**, all basal, are long and very slender, never standing erect. The slender flower stalk is about 18 inches high but may be as much as 2 feet when the plant is protected by bushes. The light blue to lavender **flowers** are ¾ to 1 inches long with a spread of the petals of about the same length. These flowers are borne in a cluster at the top of the stem. The three petals and three sepals are so near the same that the flower has the appearance of having six petals.

FLOWERING Spring.

ELEVATION Up to 5,000 feet (usually lower).

WHERE FOUND Occasional, especially on open, disturbed ground.

NOTE Also called "covena," or "grassnuts." The flowers of this species are conspicuous on the mesas and open slopes in early spring. The bulbs were eaten by the Pima and Papago Indians.

ABRAHAM ROMERO

ASTERACEAE Aster Family

COULTER'S BRICKELLBUSH

Brickellia coulteri A. Gray

DESCRIPTION Perennial herb, 8 to 15 inches high, branched throughout, the **stems** slender, whitish, glabrous except the upper ones which are sparsely woolly. **Leaves** ½ to ¾ inches long, the blades small, ovate-acute, yellow-green and sparsely short rough woolly. **Flower heads** ¼ to ⅜ inches long, bracts of several sizes, overlapping, sparsely hairy and with three purple lines running through the center of each. **Flowers** small, purplish, pappus abundant, white.

FLOWERING Spring, summer.

ELEVATION Up to 4,000 feet.

WHERE FOUND Common on dry rocky slopes and in canyons.

ZACH E PLANTS

ASTERACEAE Aster Family

SPREADING FLEABANE
Erigeron divergens Torr. & A. Gray

DESCRIPTION Slender annual, 6 to 15 inches high, with several branches from the base. **Leaves** of two kinds, the upper ones small and narrow, the lower ones 1 to 1½ inches long and sometimes two to three lobed. Stems and leaves gray-green and covered by long soft hairs. **Flower heads** many, to ¾ inches across, rays 50 or more, ¼ inch long, very slender, pale yellow at the base, lavender at the ends. Disk to ⅜ inches across, yellow.

FLOWERING Spring, summer.

ELEVATION Up to 9,000 feet.

WHERE FOUND Common on dry rocky slopes and mesas, open pine woods.

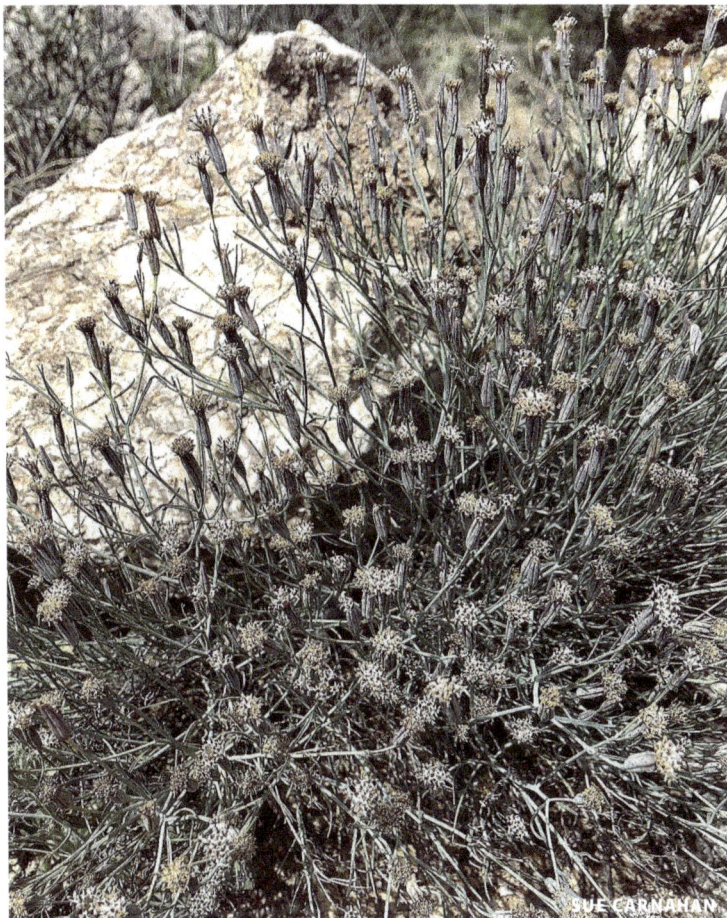

SUE CARNAHAN

ASTERACEAE Aster Family

ODORA, SLENDER PORELEAF
Porophyllum gracile Benth.

DESCRIPTION Shrub, 18 to 24 inches tall, branched throughout, stems slender, the lower ones with a gray shredding bark, the upper ones green, shallowly grooved. **Leaves** ¾ to 1 inches long, $1/16$ inch wide, very narrowly linear, dark green. **Flower heads** many, ½ to ⅝ inches long, involucre ¼ inch across, of about six rather heavy purple bracts, each bearing several oblong purple glands. **Flowers** small, rather pale purple or whitish.

FLOWERING Spring to fall.

ELEVATION Up to 4,000 feet.

WHERE FOUND Common on dry rocky slopes and in canyons.

NOTE The stems and leaves have a very unpleasant odor when crushed.

MARK POLLOCK

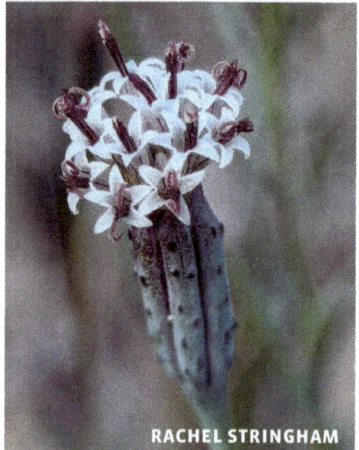
RACHEL STRINGHAM

ASTERACEAE Aster Family

WIRE-LETTUCE
Stephanomeria pauciflora
  (Torr.) A.Nelson

DESCRIPTION Shrub, 12 to 18 inches high, branched throughout, the branches slender and brittle. **Leaves** few, linear, ¼ to ⅜ inches long, dull green. **Flower heads** ½ inch across, pink or lavender-pink, composed of six to seven short rays. The pappus is densely plumose at the ends but not at the base where it has a tan color. The flowers are few and not conspicuous.

FLOWERING Summer to fall.

ELEVATION Up to 7,000 feet.

WHERE FOUND Occasional on dry plains, mesas, and slopes.

NOTE The Hopi Indians, according to one authority, apply the plant both externally and internally to stimulate milk flow in women.

DUTZA K.

ASTERACEAE Aster Family

DESERT AMERICAN-ASTER

Symphyotrichum expansum (Poepp. ex Spreng.) G.L. Nesom

SYNONYM *Aster exilis* Ell.

DESCRIPTION Coarse annual, 2 to 4 feet high, one stem from the base, this with many spreading branches above. **Leaves** 1½ to 2½ inches long, narrowly linear-acuminate, yellow-green. **Flower heads** numerous, ¼ inches across, involucral bracts in several rows, very slender. **Ray flowers** about 20, barely 1/16 inch long, pink-lavender to white; disk ⅛ inch across, pinkish. Pappus of many white hairs.

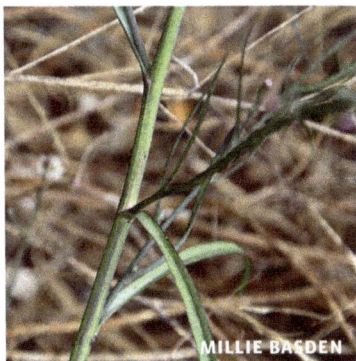

MILLIE BASDEN

FLOWERING Fall.

ELEVATION Up to 4,000 feet.

WHERE FOUND Common in moist soil along streams and ditches.

HARRIER

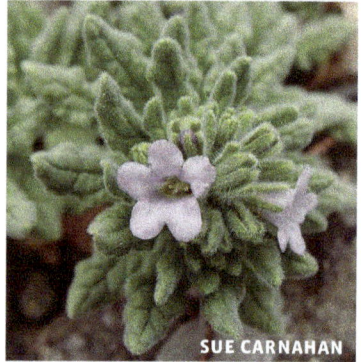

SUE CARNAHAN

BORAGINACEAE Borage Family

FIDDLELEAF

Nama hispida A. Gray

DESCRIPTION The **stems** procumbent and spreading from the center of the plant. The plants are usually quite small. **Leaves** to ¾ inch long, narrowly oblanceolate, finely hispid. **Flowers** ½ inch across, purple, corolla tube campanulate, the lobes round.

FLOWERING Spring.

ELEVATION Up to 5,000 feet.

WHERE FOUND Common on dry plains and mesas, usually in sandy soil.

NOTE Sometimes placed in its own Fiddleleaf Family (Namaceae).

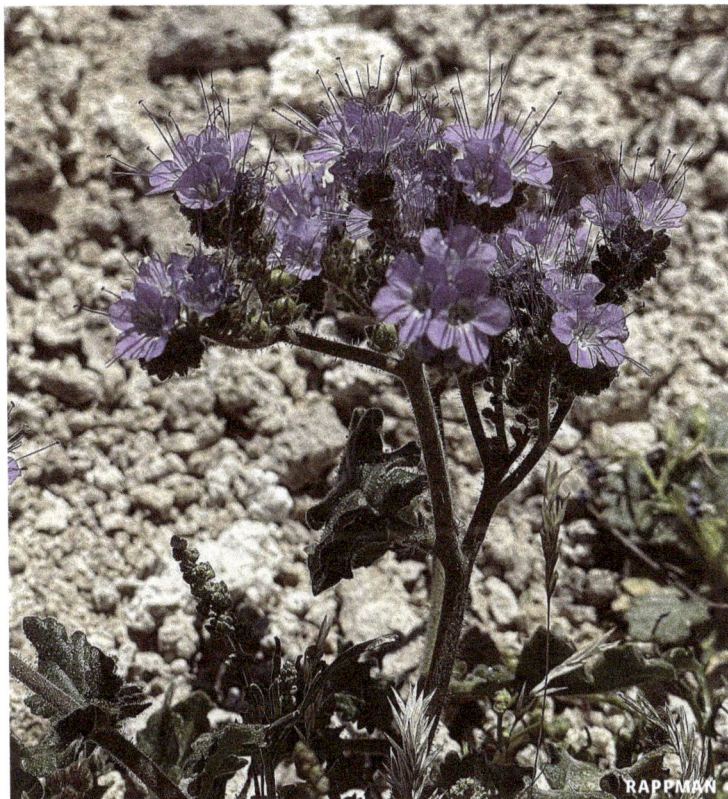

BORAGINACEAE Borage Family

NOTCH-LEAF SCORPIONWEED

Phacelia crenulata Torr. ex S. Watson

DESCRIPTION Growing tallest when in shrubs which protect it, 12 to 18 inches high. **Leaves** 1½ to 3 inches long, pinnately dissected to the midrib, the lobes with the margins coarsely toothed. Stems and leaves hispid. **Inflorescence** a helicoid cyme, the flowers all borne on one side of the stem. **Flowers** ½ inch across, corolla tube campanulate, yellowish, the upper part and the rounded lobes a deep blue. The long blue filaments surmounted by the dark yellow anthers are very prominent. Calyx lobes and fruits strongly hispid.

FLOWERING Spring.

ELEVATION Up to 4,000 feet.

WHERE FOUND Very common on plains, mesas, and foothills.

NOTE Sometimes called "wild-heliotrope," Arizona's most abundant species of *Phacelia*, very conspicuous in spring with its rich violet-purple flowers, the plant glandular-viscid and with an unpleasant, somewhat onion-like odor.

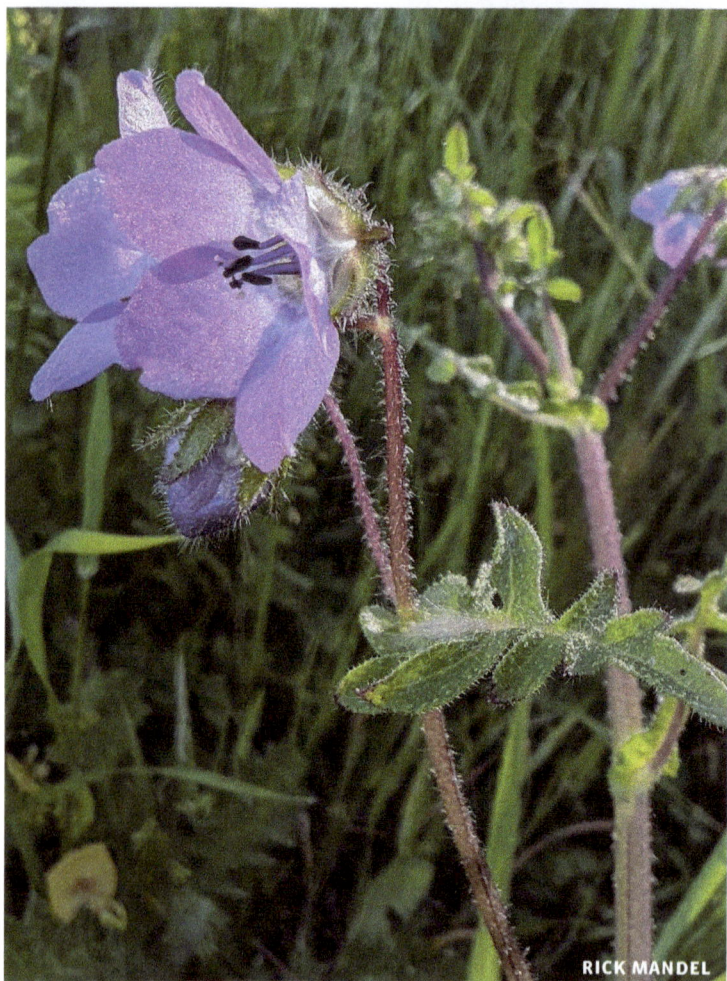

RICK MANDEL

BORAGINACEAE Borage Family

BLUE FIESTA-FLOWER

Pholistoma auritum (Lindl.) Lilja

DESCRIPTION Very slender annual, 12 to 24 inches tall, somewhat trailing, the **stems** very weak. **Leaves** ¾ to 1 inch long, pinnate, the lobes small and entire. Leaves and stems sparsely hispid. **Flowers** solitary on long slender pedicels. Flowers ⅜ inch across, medium blue. Calyx lobes ¼ inches long, acute, strongly hispid, the sinus appendages also acute and hispid.

FLOWERING Spring.

ELEVATION Up to 3,000 feet.

HABITAT Common in damp rocky places in streambeds, rocky slopes.

JACK BYRLEY

CACTACEAE Cactus Family

GRAHAM'S NIPPLE CACTUS
Cochemiea grahamii (Engelm.) Doweld

SYNONYM *Mammillaria grahamii* Engelm.
DESCRIPTION Globose to cylindric, simple or budding either at the base or near it, often tufted but in small clusters. **Stems** 6 inches high, tuberculate, the tubercles small, corky when old, the axils naked. **Radial spines** fifteen to twenty, ⅜ to ½ inches long, white, spreading, sometimes with dark tips. **Central spines** one to three, dark, when more than one the lower one stouter, often 1 inches long, hooked. **Flowers** from near the top of the plant, ¾ to 1 inches long, pink-purple. **Fruit** clavate, ¾ to 1 inches long, scarlet; seeds black, pitted, globose.
FLOWERING Spring to summer.
ELEVATION Up to 4,500 feet.
WHERE FOUND Occasional in desert scrub, soils silty, sandy, gravelly or rocky.

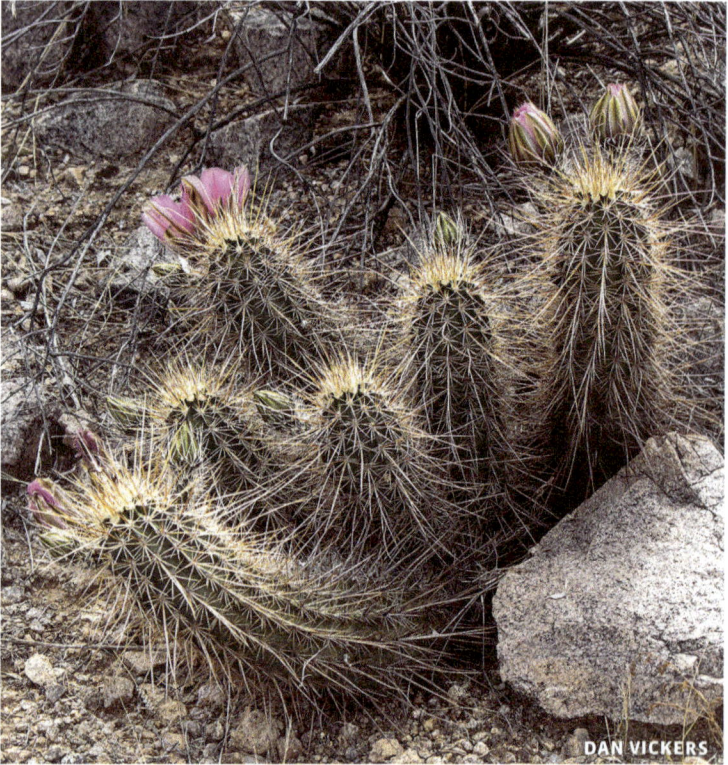

DAN VICKERS

CACTACEAE Cactus Family

HEDGEHOG CACTUS, STRAWBERRY CACTUS

Echinocereus engelmannii (Parry ex Engelm.) Lem.

DESCRIPTION Tufted, forming large clumps; **stems** erect or nearly so, cylindric, 6 to 15 inches long, 2 to 2½ inches in diameter. **Ribs** eleven to fourteen, low, obtuse; areoles large, nearly circular; **radial spines** about ten, stiff, appressed, ½ inch long. The **central spines** five to six, very stout, more or less curved and twisted, about 2½ inches long, yellowish to brown. **Flowers** 2 to 3 inches long, even broader when fully epanded, pink-purple in color. **Fruit** spiny, about 1¼ inches long; seeds numerous, black.

FLOWERING Spring.

ELEVATION Up to 5,000 feet.

WHERE FOUND Common plant of desert flats.

NOTE A variable species; ordinarily spines of more than one color occur at the same areole, and they may be white, brown, black, or yellow, opaque or rarely translucent.

RACHEL STRINGHAM

CARYOPHYLLACEAE Pink Family

SLEEPY CATCHFLY

Silene antirrhina L.

DESCRIPTION Slender, 10 to 12 inches high, somewhat branched through-out, the **stems** light green and on each internode bearing a dark green-brown sticky band about ½ to ¾ inches long. **Leaves** opposite, slender, 1 to 2 inches long. **Flowers** pink. Seeds small, black and very numerous.

FLOWERING Spring.

ELEVATION Up to 6,000 feet.

WHERE FOUND Occasional in open areas.

IAN LANE

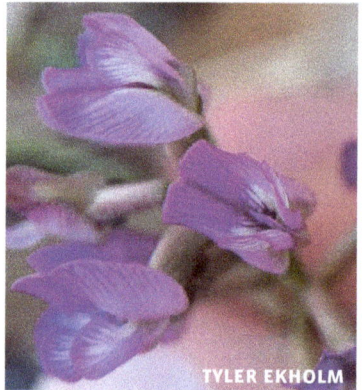
TYLER EKHOLM

FABACEAE Pea Family

SHEEP MILKVETCH
Astragalus nothoxys A. Gray

DESCRIPTION Low growing, 3 to 5 inches high; **stems** spreading, 5 to 10 inches long. **Leaves** 1 to 1½ inches long, odd pinnate, four to six pairs, the leaflets to ¼ inches long, ovate-elliptical, gray-green. The leaves and stems sparsely short strigose. **Flowers** about ¼ inch long, lavender with dark purple stripes. The keel is sometimes pale yellow. **Pods** ½ inch long, laterally flattened and curved, sparsely short strigose.

FLOWERING Spring.

ELEVATION Up to 6,000 feet.

WHERE FOUND Common on slopes and mesas, often with oaks.

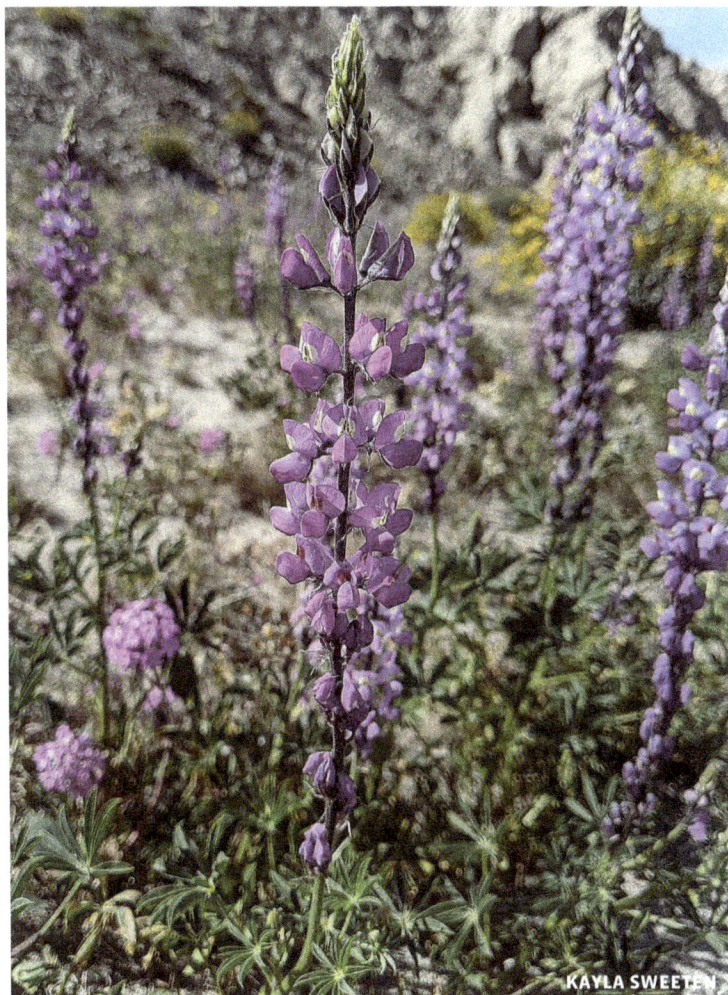

KAYLA SWEETEN

FABACEAE Pea Family

ARIZONA LUPINE

Lupinus arizonicus (S. Watson) S. Watson

DESCRIPTION Coarse, 6 to 12 inches high, one to several stems from the base, these yellow-green, sparsely long hirsute. **Leaves** 2 to 3 inches long, palmate, six to eight leaflets; **leaflets** ¾ to 1 inches long, ³/₁₆ to 5/16 inches wide, oblanceolate, medium green, sparsely long hirsute. **Flowers** ⅜ inch long, blue-lavender. **Pods** ¾ inch long, ³/₁₆ inches wide, straight, yellow-green.

FLOWERING Spring.

ELEVATION Up to 3,000 feet.

WHERE FOUND Common in sandy washes.

KATE CROWELL-WALKER

FABACEAE Pea Family

MOJAVE LUPINE

Lupinus sparsiflorus Benth.

DESCRIPTION Several stems from the base, 12 to 15 inches tall. **Leaves** 2 to 3 inches long, palmate; **leaflets** 6, these ¾ to 1¼ inches long, ⅛ inch wide, very narrowly oblanceolate-acute, medium green. Stems and leaves very sparsely long strigose. **Flowers** ½ inch long, petals large, dark blue. Pods straight, strigose.

FLOWERING Spring.

ELEVATION 4,500 feet or lower.

WHERE FOUND Common on mesas and foothills, preferring sandy soil.

NOTE In favorable springs this handsome lupine colors extensive areas with the rich violet of its flowers.

KEN BOSMA

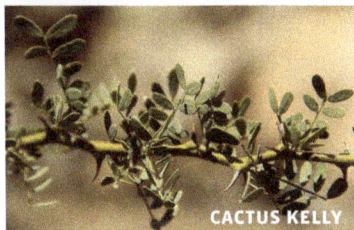
CACTUS KELLY

FABACEAE Pea Family

DESERT-IRONWOOD
Olneya tesota A. Gray

DESCRIPTION Tree, 10 to 15 feet high, branched throughout, **stems** gray-green, with two long straight spines at the base of each leaf cluster. **Leaves** 1½ to 2 inches long, leaflets six to eight pairs, ⅜ inch long, oblong-obovate to oblong, minutely puberulent, light blue-green. **Flowers** in clusters, ¾ inch long, lavender. **Pods** 3 to 3½ inches long, pubescent; **seeds** few, very apparent in the pods.

FLOWERING Spring.

ELEVATION Up to 2,500 feet

WHERE FOUND Common along washes in the foothills.

NOTE Known commonly in Arizona as "ironwood," or "palo-de-hierro". A desert tree with very handsome flowers, limited to warm locations, and for this reason was used as an indicator in selecting sites for citrus orchards. The foliage is evergreen except in very cold winters. The wood is brittle, hard, remarkably heavy, and burns very slowly, making good coals. The ironwood has been used so extensively for firewood that it is unusual to find a large tree that has not been cut back to the

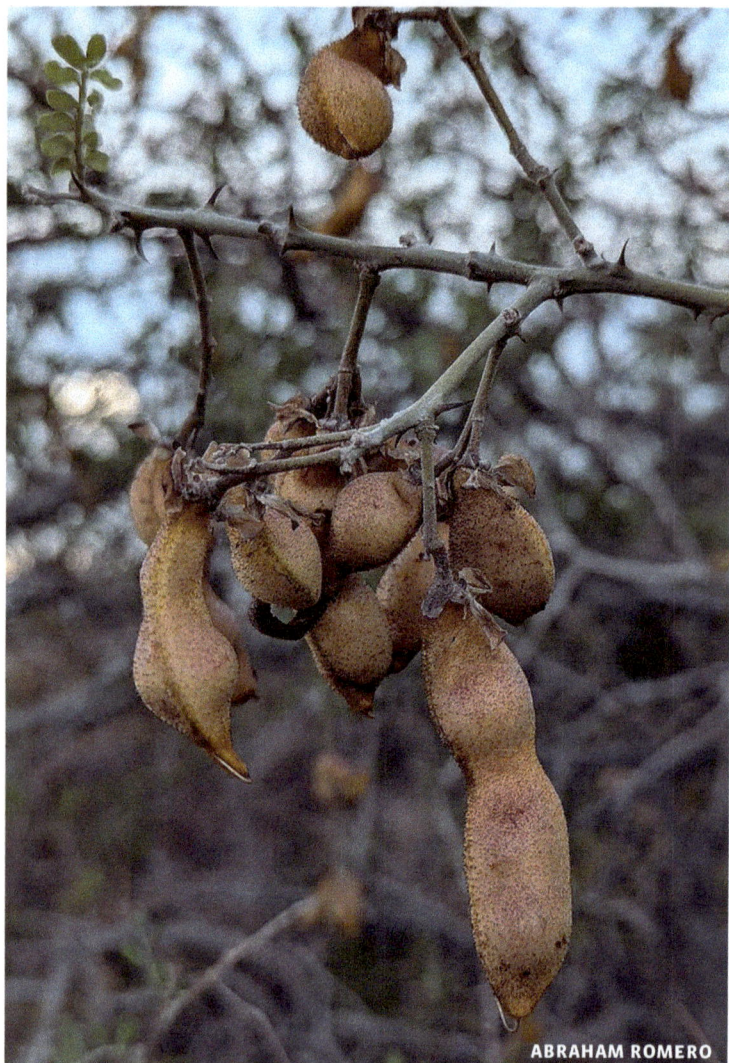

ABRAHAM ROMERO

stump. The wood was used by the Indians for arrowheads and for tool handles. Experiments have been made to utilize it commercially, but it is too hard for ordinary woodworking tools. The hard sharp spines of the branches do not prevent desert-bred horses from eating the foliage with evident relish. The seeds are an important food of desert animals and formerly were eaten parched by the Pima Indians.

JACQUELINE WORTHINGTON

SAM KIESCHNICK

GERANIACEAE Geranium Family

RED-STEM STORK'S-BILL
Erodium cicutarium (L.) L'Hér.

DESCRIPTION Low growing, 4 to 6 inches high; branched from the base, the branches spreading on the ground. **Leaves** ½ to 1¼ inches long, pinnately divided, the lobes again finely divided, green. Stems and leaves sparsely long soft hairy. **Flowers** pink-lavender; star-like. Carpel tails 1½ inch long, sparsely hairy, curling back from the central column when mature.

FLOWERING Spring.

ELEVATION Up to 7,000 feet.

WHERE FOUND Common on plains, mesas, mountains.

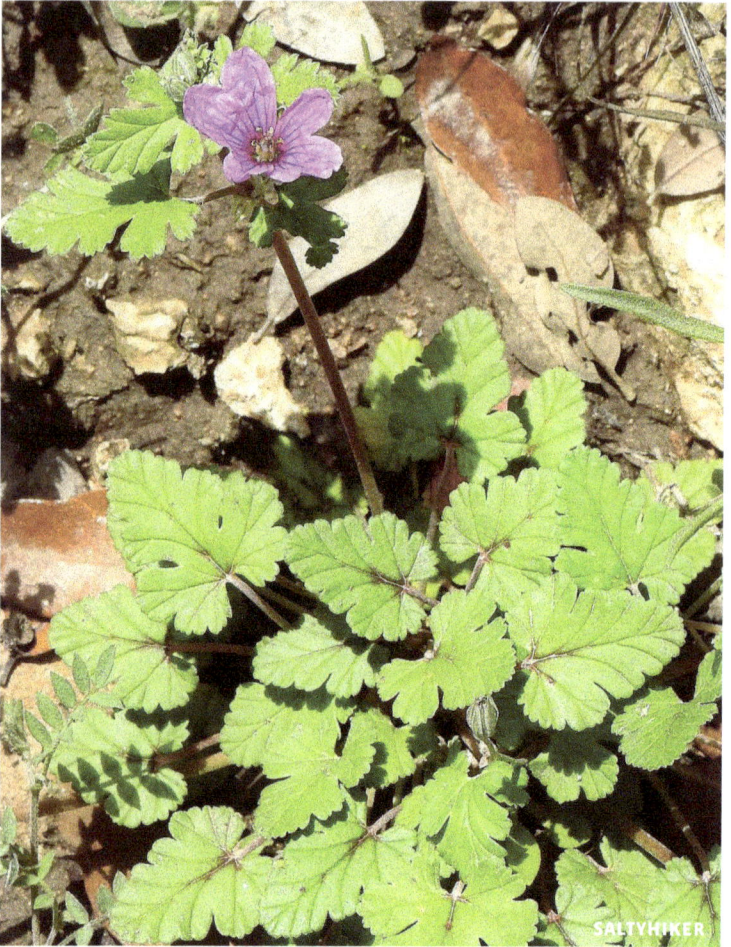

GERANIACEAE Geranium Family

TEXAS STORK'S-BILL

Erodium texanum A. Gray

DESCRIPTION Low growing 4 to 6 inches tall, the branches spreading 1 to 2 feet. **Leaves** opposite, 1 to 2 inches long, palmately divided into three sections, these in turn shallowly lobed at the tips. Leaves and stems minutely pubescent. **Flowers** purple. Carpel tails as long as 2 inches, each curling back leaving the slender erect column in the center.

FLOWERING Spring.

ELEVATION Up to about 3,000 feet.

WHERE FOUND Common on plains and mesas.

ALEXANDER WENTWORTH

HELIOTROPIACEAE
Heliotrope Family

ALKALI HELIOTROPE
Heliotropium curassavicum L.

AMANDA FISHER

DESCRIPTION Plants low growing, branched somewhat, spreading. The **leaves** and **stems** thick, smooth, and gray-green because of the bloom which covers them. **Flowers** crowded on the upper surfaces of the coiled stems. The small, pale lavender to whitish flowers have a pale yellow center. The **fruit** consists of four nutlets.

FLOWERING Spring.

WHERE FOUND Salty or alkaline soil; not abundant around Phoenix.

NOTE Formerly placed in Borage Family (Boraginaceae). These plants are sometimes known as "Chinese-pusley" and "quailplant." It is stated that the dried root, finely powdered, was applied to sores and wounds by the Pima Indians.

RACHEL STRINGHAM

KRAMERIACEAE Ratany Family

WHITE RATANY
Krameria bicolor S. Watson

SYNONYM *Krameria grayi* Rose & Painter

DESCRIPTION Low shrub, 1½ to 2 feet high, branched throughout, the branches with a short silver-gray pubescence. **Leaves** ¼ to ½ inches long, slender, light green, also pubescent. **Flowers** ¾ inch across, irregular, the five sepals being very irregular and deep crimson, the petals reduced to three small erect upper appendages and two small lower ones in the center of the flower; stamens three. **Fruit** an indehiscent globose pod covered with long red spines.

FLOWERING Spring to fall.

ELEVATION Up to 4,000 feet.

WHERE FOUND Common on dry, rocky and sandy soils in desert scrub.

SMECKERT

LAMIACEAE Mint Family

DESERT-LAVENDER

Condea emoryi (Torr.) Harley & J.F.B. Pastore

SYNONYM *Hyptis emoryi* Torr.

DESCRIPTION Shrub, branched throughout, especially at the base. The branches are opposite and come out nearly at right angles to the main stems. **Leaves** ½ to 1 inches long, ¾ inch wide, cordate, margins shallowly toothed, texture heavy, rugose. Stems and leaves densely short, gray hairy. **Flowers** ⅜ inch long, blue-lavender, not showy; calyx densely gray-woolly.

FLOWERING Spring, or almost throughout the year at lower elevations.

ELEVATION Up to 5,000 feet (usually lower).

WHERE FOUND Common on dry rocky slopes and canyons.

NOTE The plant is browsed to a limited extent by livestock. The seeds, like those of certain species of *Salvia,* are used in Mexico as food under the name "chia."

STEVE MATSON

LAMIACEAE Mint Family

CALIFORNIA SAGE, CHIA
Salvia columbariae Benth.

DESCRIPTION Several stems from the base; plants short gray-hairy, gray-green in color.. **Leaves** all basal, 2 to 4 inches long, pinnately divided to the mid-rib, the lobes irregularly, shallowly lobed and rugose. **Stems** four-sided, erect and bearing at the top usually two ball-like clusters of flowers subtended by many spiny bracts. **Flowers** small, $3/16$ inch across, blue to blue-lavender and quite inconspicuous because of the long spine-tipped calyx lobes.

FLOWERING Spring.

ELEVATION Up to 3,500 feet.

WHERE FOUND Common in sandy washes.

NOTE One of the species known as "chia." The seeds were utilized by the Indians to make pinole and also mucilaginous poultices. A beverage prepared from the seeds was popular with the Pima Indians. The seeds of other species known as chia are extensively used in Mexico for similar purposes.

SUE CARNAHAN

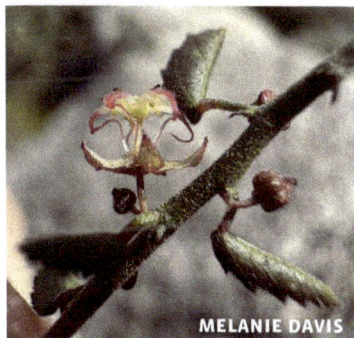

MELANIE DAVIS

MALVACEAE Mallow Family

CALIFORNIA AYENIA
Ayenia compacta Rose

DESCRIPTION Low growing, 6 to 12 inches high; **stems** slender, spreading; **leaves** ½ to 1 inch long, narrowly oblong, the margins toothed; upper surface red-brown, lower surface green; stems and leaves covered with a minute pubescence. **Flowers** pale pink, inconspicuous. **Fruit** a globose capsule, covered with small red prickles.

FLOWERING Summer.

ELEVATION Up to 4,000 feet.

WHERE FOUND Occasional, mostly in sandy washes.

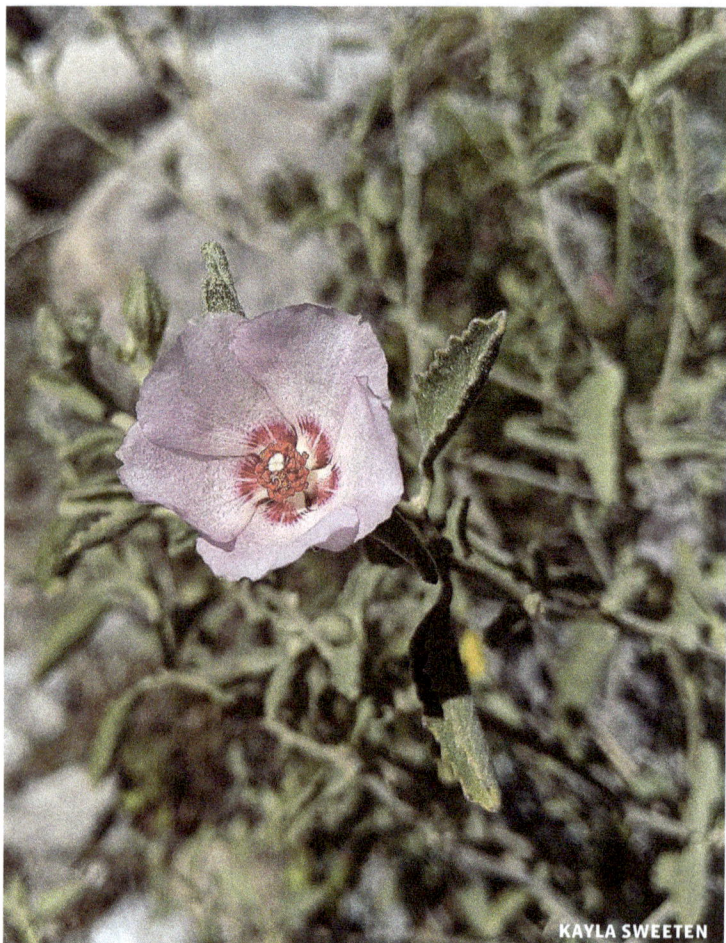

KAYLA SWEETEN

MALVACEAE Mallow Family

ROCK HIBISCUS
Hibiscus denudatus Benth.

DESCRIPTION Low growing, 12 to 18 inches high, with a few **stems** from the base, these and the few ovate leaves are very dull gray-green and quite unattractive. The **leaves** are 1 to 1½ inches long. **Flowers** showy, 1½ inches across, lavender with a dark purple center and stamen column. The **fruits** are very conspicuous when they split open to disperse the seeds as the carpels stay on the plants. The dark seeds each bear a fringe of short soft hairs around the sides.

FLOWERING Spring.

ELEVATION Up to about 2,000 feet.

WHERE FOUND Sandy washes, not common.

MATT BERGER

NYCTAGINACEAE Four-O'clock Family

DESERT SAND VERBENA
Abronia villosa S. Watson

DESCRIPTION Low growing, **stems** trailing 1 to 3 feet, densely villous or hairy. **Leaves** ovate or oblong, yellow-green. **Flowers** borne in flat topped clusters similar to those of the true Verbena. Individual flowers pink-lavender, ¼ inch across.

FLOWERING Spring to summer.

ELEVATION Up to about 1,500 feet.

WHERE FOUND Common in dry sandy places, in spring may cover sandy riverbanks with a blaze of bright color.

RACHEL STRINGHAM

NYCTAGINACEAE Four-O'clock Family

TRAILING WINDMILLS
Allionia incarnata L.

DESCRIPTION Prostrate perennial, **stems** spreading 5 to 12 inches, short glandular hairy and sticky. **Leaves** unequal, the larger one ¾ to 1 inches long, oblong-ovate, the margins nearly entire and deeply undulate, both surfaces gray-green and hairy with glandular hairs. **Flowers** ½ inch across, each composed of three individual flowers in an involucre. The flowers are rose-purple and the stamen anthers bright yellow.

FLOWERING Spring to fall.

ELEVATION Up to 6,000 feet.

WHERE FOUND A conspicuous, common plant on open plains, mesas, and slopes. The plant is always covered with sand which sticks to the glandular hairs.

ANTHONY BATISTA

NYCTAGINACEAE Four-O'clock Family

SCARLET SPIDERLING

Boerhavia coccinea Mill.

DESCRIPTION Low growing perennial, the **stems** trailing on the ground. **Leaves** mostly basal, only a few smaller ones being found near the ends of the branches; leaves about 1½ inches long, irregular, ovate and somewhat sticky. The long irregularily branched stems bear small clusters of tiny red-purple **flowers** about ⅛ inch across.

FLOWERING Summer.

ELEVATION Up to 5,500 feet.

WHERE FOUND Very common in sand washes, also on roadsides and in fields.

NOTE With its long trailing stems and viscid herbage, this is sometimes an annoying weed in gardens.

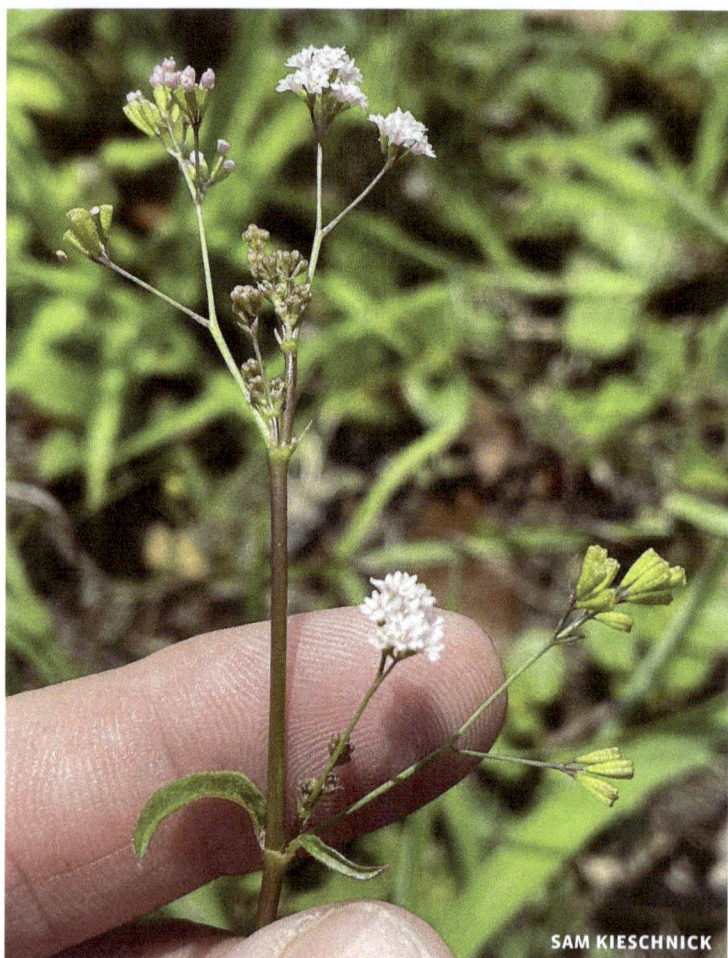

SAM KIESCHNICK

NYCTAGINACEAE Four O'clock Family

ERECT SPIDERLING
Boerhavia erecta L.

DESCRIPTION Slender, 6 to 15 inches high, **stems** minutely puberulent. **Leaves** opposite, of different sizes, the larger one about 1 to 1½ inches long, ¼ to ⅜ inches wide, dark red on the upper surface, light green below. **Flowers** tiny, ³/16 inch across, pale pink to nearly white.

FLOWERING Summer to fall.

ELEVATION Up to 5,500 feet.

WHERE FOUND Common in sandy open places, often where disturbed.

MOSTBITTERN

OROBANCHACEAE Broom-Rape Family

PURPLE INDIAN-PAINTBRUSH
Castilleja exserta (A. Heller) Chuang & Heckard

SYNONYM *Orthocarpus purpurascens* Benth.

DESCRIPTION Low growing, 3 to 8 inches high, usually only one stem from the base, this with many long slender green **leaves** with a sparse pubescence. Near the top of the stem the leaves are larger, deeply parted into acuminate divisions and have the tips tinged with red-lavender. The leaves in the flower head are all red-lavender. The **flowers** are about ½ inch long, have a large bulbous lower lip and a rather long, narrow, fuzzy upper lip. The stamens are yellow. The whole plant very closely resembles a soft plume.

FLOWERING Spring.

ELEVATION Up to 3,500 feet.

WHERE FOUND Occasional on open mesas and slopes.

NOTE Also known by the Spanish name "escobita." In favorable seasons, extensive areas are bright purple with the flowers of this plant, which is grazed by cattle and sheep.

MELANIE DAVIS

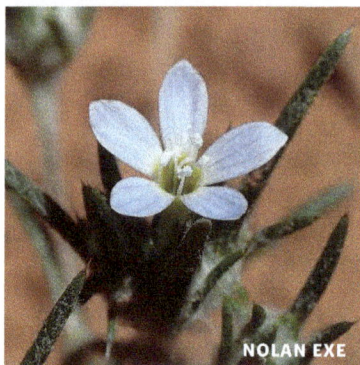

NOLAN EXE

POLEMONIACEAE Phlox Family

DESERT WOOLLYSTAR

Eriastrum diffusum (A. Gray) Mason

DESCRIPTION Much branched throughout, 4 to 6 inches high, the stems usually dark red-brown, very sparsely hairy. **Leaves** ½ to ¾ inches long, slender, reddish. **Flowers** ½ inch long, corolla tubular, pale yellow, corolla lobes ³/₁₆ to ¼ inches long, wide spreading, blue or white. The flowers are like tiny stars. The calyx is densely white-woolly.

FLOWERING Spring.

ELEVATION Up to 4,000 feet.

WHERE FOUND Common on plains and mesas.

RICK MANDEL

POLEMONIACEAE Phlox Family

ROSY GILIA
Gilia sinuata Dougl. ex Benth.

DESCRIPTION Low growing, 3 to 15 inches tall, very slender. **Leaves** few and finely dissected, those at the base of the plant larger than the others. **Flowers** about ¼ inch across, pale lavender-blue to nearly white. The five spreading pointed petals giving the flower the appearance of a tiny star.

FLOWERING Spring.

ELEVATION Up to 7,000 feet (usually much lower).

WHERE FOUND Very common, usually in open sandy places.

STAN SHEBS

EGRETEN

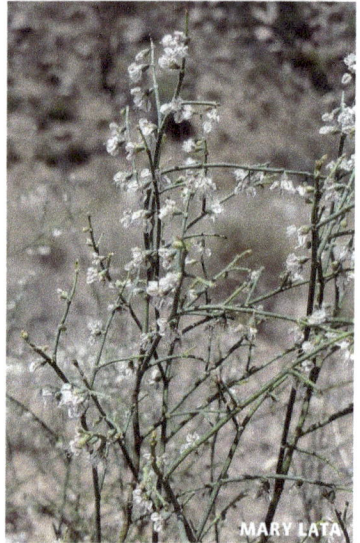
MARY LATA

POLYGONACEAE
Buckwheat Family

FLATCROWN BUCKWHEAT

Eriogonum deflexum Torr.

DESCRIPTION Annual, 1 to 2 ft. high, one to several **stems** from a basal rosette of heavily felted, nearly round, gray **leaves** 1 to 2 inches long. The stems branch in three's usually about 2 inches from the ground, these in turn branch until the plant resembles a densely branched tree with no leaves. The **flowers** are numerous, borne in clusters throughout at the ends of the tiny branchlets. The flowers are about ¼ inch across, very pale pink to nearly white. The top of the plant is flat.

FLOWERING Summer to fall.

ELEVATION Up to 4,500 feet.

WHERE FOUND Common on roadsides, sandy desert scrub.

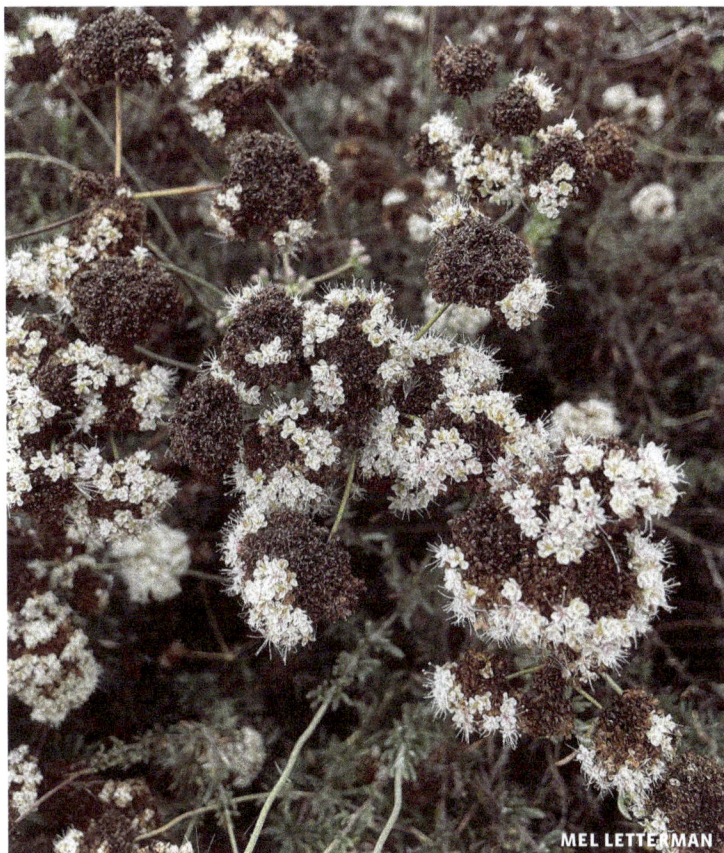

MEL LETTERMAN

POLYGONACEAE Buckwheat Family

CALIFORNIA BUCKWHEAT
Eriogonum fasciculatum Benth.

DESCRIPTION Shrub, 2 to 3 feet high, bark on lower branches red-brown and shredding, on the younger erect branches a silver-gray. **Leaves** ½ to ¾ inches long, oblanceolate, gray-green with an inconspicuous matted pubescence on the under side. The leaves are borne in small clusters along the stems. The **flowers** are in clusters arranged at the ends of small spreading branchlets ¼ to 1 inches long, which are borne at the apices of the long, slender, leafless, flower stalks. The flowers are dark pink when in bud and pale pink to nearly white and very pretty when in full bloom. Long after flowering the old stalks and flowers remain on the bushes. These stalks are dark red-brown and very conspicuous.

FLOWERING Spring.

ELEVATION 2,000 to 4,500 feet.

WHERE FOUND Common on dry rocky slopes.

POLYGONACEAE Buckwheat Family

THOMAS' WILD BUCKWHEAT
Eriogonum thomasii Torr.

DESCRIPTION Annual, 6 to 10 inches tall, one stem from a basal rosette of very small gray, woolly **leaves**. The stem branches 1 to 1½ inch from the ground into four, five or six branches which branch again until the plant is densely branched throughout. **Flowers** very numerous, borne in clusters at the ends of tiny branchlets. Flowers about ⅛ inch wide, pale pink. In some places the plants cover the ground with a lacy covering of these tiny flowers.

FLOWERING Spring.

ELEVATION Up to 3,000 feet.

WHERE FOUND Sandy soil, common.

NOTE This is one of the most delicate of our native flowers.

KLICKLO

RANUNCULACEAE Buttercup Family

OCEAN-BLUE LARKSPUR
Delphinium parishii A. Gray

DESCRIPTION Rather coarse, 2 to 3 feet high. The **lower leaves** are 2 to 3 inches across, slashed nearly to the center, the **upper leaves** are small and inconspicuous. **Flowers** 1 to 1½ inches across and an attractive shade of light blue. The flowers are borne in a raceme 8 to 10 incheslong.

FLOWERING Spring.

ELEVATION Up to 3,500 feet.

WHERE FOUND Occasional on rocky knolls and desert mesas.

NOTE This is the most xerophytic of North American larkspurs. The reduction in vegetative parts enables the plants to withstand withering heat or prolonged droughts.

RODOLFO VILLARREAL

SUE CARNAHAN

SOLANACEAE Potato Family

WILD PETUNIA

Calibrachoa parviflora (Juss.) D'Arcy

SYNONYM *Petunia parviflora* Juss.

DESCRIPTION Low growing annual, 3 to 5 inches high, the stems trailing 5 to 6 inches. **Leaves** linear, small, entire, light green. **Flowers** pale lavender or purple, ¼ inch across, solitary, axillary. **Fruit** an ovoid capsule about ¼ inch long, surpassed by the calyx lobes.

FLOWERING Spring.

ELEVATION Up to 5,000 feet.

WHERE FOUND Common along stream beds.

NOTE The flowers are like those of the cultivated petunia except in size.

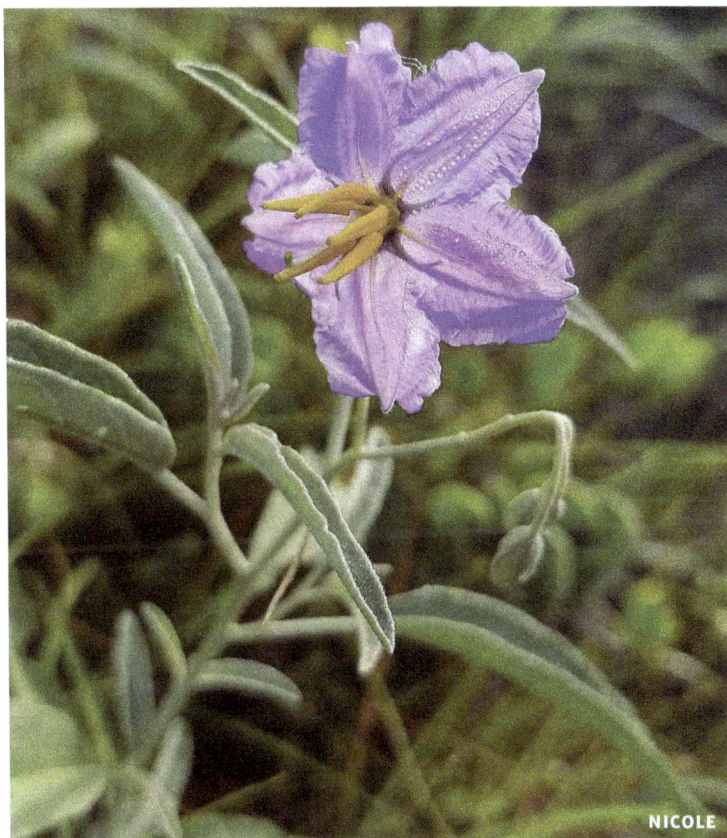

NICOLE

SOLANACEAE Potato Family

SILVERLEAF NIGHTSHADE

Solanum elaeagnifolium Cav.

DESCRIPTION Annual, 1 to 3 feet high, branched at the base, the stems rather heavy and tough. **Leaves** 2 to 3 inches long, narrow, the margins deeply scalloped or undulate. Leaves with a silvery white pubescence. The stems and veins of the leaves bear long, slender, brown spines. **Flowers** numerous, 1 inch across, the petals united, forming a saucer-shaped corolla; petals blue-lavender. The five slender, yellow anthers stand erect in the center of the flower.

FLOWERING Spring to fall.

ELEVATION Up to 5,500 feet.

WHERE FOUND Common along roadsides and in fields.

NOTE A troublesome weed in irrigated land, and difficult and expensive to eradicate. The crushed berries are added to milk by the Pima Indians in making cheese. A protein-digesting enzyme, resembling papain, is found in this plant.

SUE CARNAHAN

ACANTHACEAE Acanthus Family

HUMMINGBIRD-BUSH, JUPAROSA

Justicia californica (Benth.) D.N. Gibson

SYNONYM *Beloperone californica* Benth.

DESCRIPTION Shrub, 2 to 3 feet high, much branched throughout, the branches slender and gray-green, covered with a soft inconspicuous pubescence. **Leaves** very few, these small, heart-shaped, medium green. **Flowers** 1¼ inches long, tubular, two-lipped, orange-red, yellow near the top of the tube.

FLOWERING Spring to fall.

ELEVATION Up to 3,500 feet.

WHERE FOUND Common on dry rocky slopes.

NOTE Called "chuparosa" in Sonora. The plant is browsed to some extent by livestock. The flowers are very attractive to hummingbirds, and it is reported that they were eaten by the Papago Indians. The name "honeysuckle" is sometimes used locally for this plant.

CHRIS ENGLISH

LOGAN BRADLEY

FOUQUIERIACEAE Ocotillo Family

OCOTILLO

Fouquieria splendens Engelm.

DESCRIPTION Six to 12 feet high with many slender stems from the base, these leafless except after rains then bearing clusters of narrow dark green **leaves** 1 to 2 inches long. At the base of each cluster of leaves is a stout gray spine. The spines are nearly at right angles to the stems. During the flowering season each branch bears at its summit an elongated cluster of bright orange-red **flowers**. The individual flowers are very glossy, the corolla is about 1 inches long and the slender stamens extend about ½ inch beyond it.

FLOWERING Spring to summer.

ELEVATION Up to 5,000 feet.

WHERE FOUND Very common on dry mesas and rocky slopes.

NOTE The well-known ocotillo of the southwest, Sometimes called "slim-wood" or "coach-whip." It is one of the oddest and most conspicuous of Arizona plants and is very attractive when in flower. The ocotillo drops its leaves as soon as the soil dries, but as quickly refoliates after a good rain, except in winter when temperatures are low.

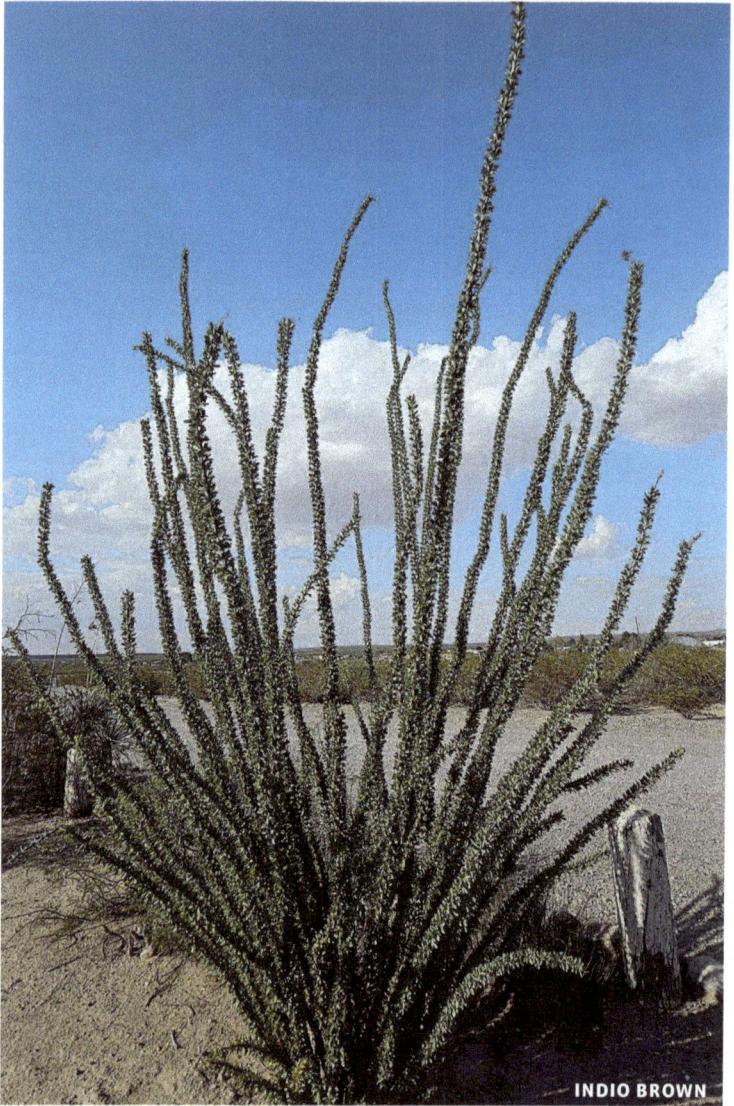

INDIO BROWN

Cuttings root readily, and it is not uncommon to see living fences or hedges of this plant. The straight, thorny stems are set in the ground thickly to build coyote-proof runs and corrals for fowl, and also were utilized by the Indians for constructing crude huts and outhouses. The Coahuila Indians of southern California were said to eat the flowers and capsules. It is reported that belt dressing of good quality is manufactured from the wax that coats the stems. The Apache Indians relieve fatigue by bathing in a decoction of the roots, and also apply the powdered root to painful swellings.

FERN WILDTRUTH

MALVACEAE Mallow Family

PELOTAZO, VELVET-LEAF MALLOW

Abutilon incanum (Link) Sweet

SYNONYM *Abutilon pringlei* Hochr.

DESCRIPTION Shrub, much branched throughout, 2 to 3 feet tall. **Leaves** small, heart-shaped, light green with a fine stellate or velvety pubescence. **Flowers** many, about ⅜ inch across, the petals rounded and pale orange, each with a reddish line down through the center.

FLOWERING Spring.

ELEVATION Up to 4,000 feet.

WHERE FOUND Common on dry slopes.

NOTE The cluster of bright yellow anthers topping a red stamen column is distinctive. Fibers extracted from the stems are reported to be used in Mexico for making rope.

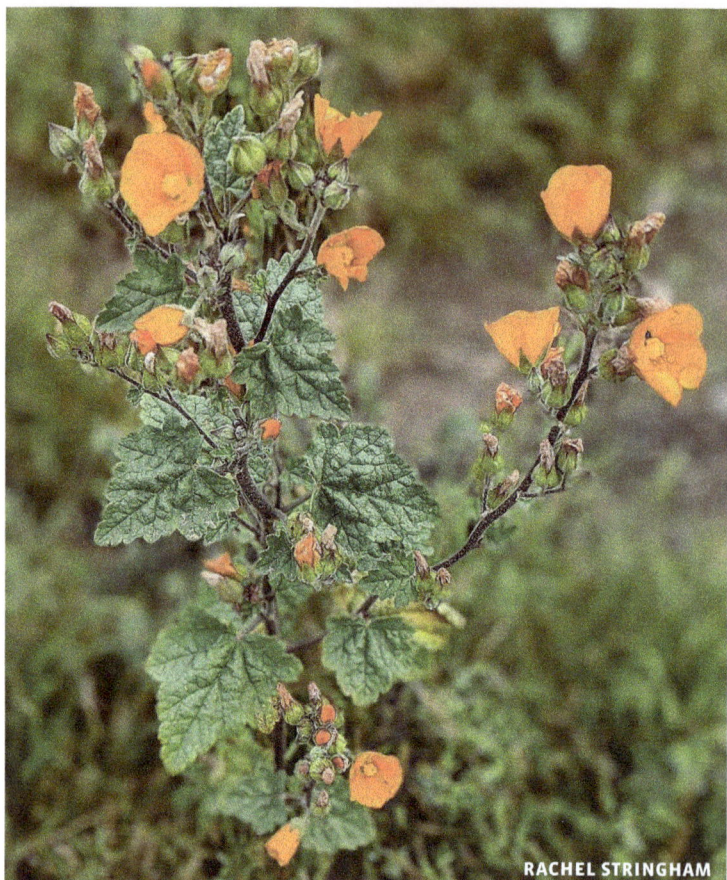

RACHEL STRINGHAM

MALVACEAE Mallow Family

COULTER'S GLOBEMALLOW
Sphaeralcea coulteri (S. Watson) A. Gray

DESCRIPTION Winter annual, 1 to 2 feet high, many branches from the base, the central ones erect, the outer ones spreading. **Leaves** 1½ to 2½ inches long, the blades obscurely two-lobed at the base and with the margins irregularly scalloped. Leaves and stems rough stellate hairy. **Flowers** many, 1 inch across, bright orange-yellow, the conspicuous stamen column with bright yellow stamens.

FLOWERING Spring.

ELEVATION Up to 2,500 feet.

WHERE FOUND Common on roadsides, fields, mesas, dry sandy desert flats.

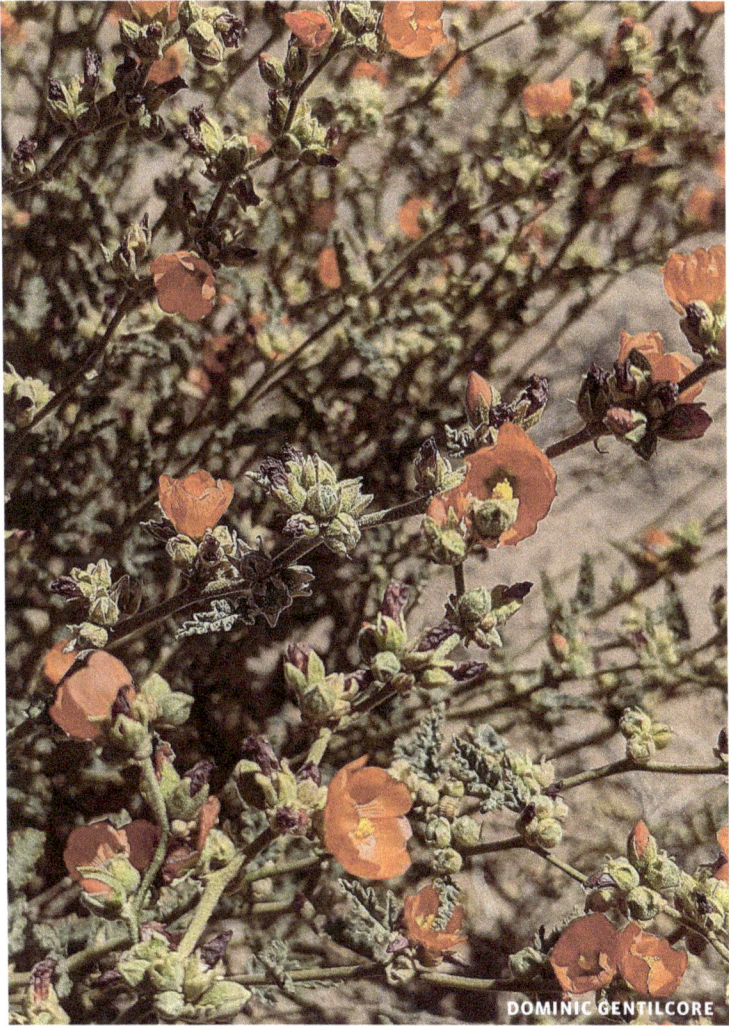

DOMINIC GENTILCORE

MALVACEAE Mallow Family

EMORY'S GLOBE-MALLOW

Sphaeralcea emoryi Torr. ex A. Gray

DESCRIPTION Perennial, somewhat woody at base, 2 to 3 feet high, many
stems from the base, these usually not branched. **Leaves** 2 to 3 inches
long, blades lobed at the base, margins irregularly scalloped. Leaves and
stems rough stellate hairy. **Flowers** very numerous, 1 to 1½ inches
across, bright orange.

FLOWERING Spring.

ELEVATION Up to about 2,500 feet.

WHERE FOUND Very common on roadsides and in fields.

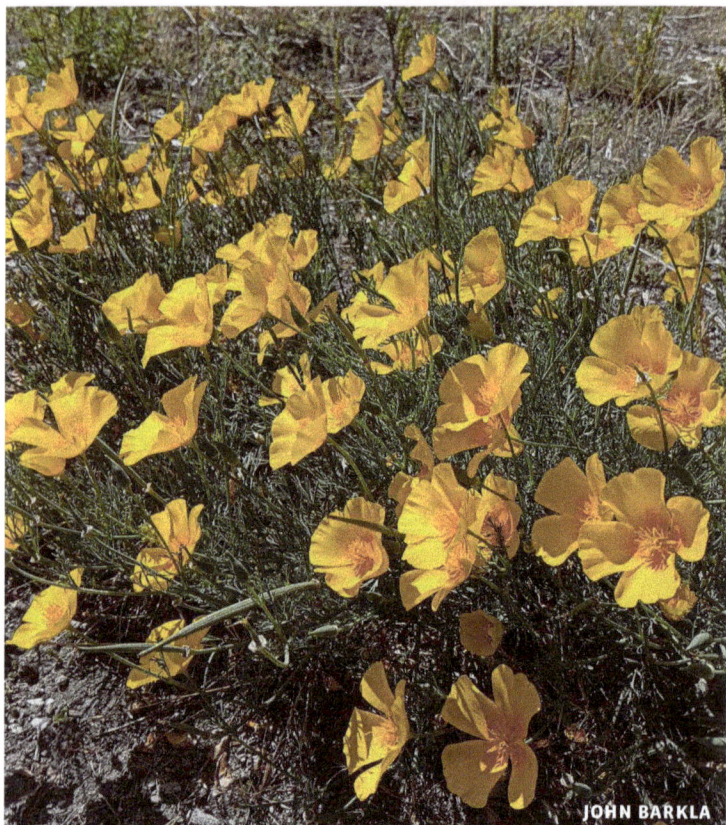

JOHN BARKLA

PAPAVERACEAE Poppy Family

CALIFORNIA POPPY
Eschscholzia californica Cham.

SYNONYM *Eschscholzia parvula* (A. Gray) Cockerell

DESCRIPTION A slender annual herb, 6 to 10 inches high. **Leaves** gray-green, finely dissected, mostly basal. Flower stalks erect and leafless. **Flowers** regular, bright orange or orange-yellow, 1 to 1½ inches across; petals four; sepals fused into a pointed cap which comes off when the flower is ready to open; stamens numerous. **Fruit** a one-celled capsule.

FLOWERING Spring.

ELEVATION Below 4,000 feet.

WHERE FOUND Common on rocky hillsides, plains, mesas, disturbed areas, roadsides and washes.

NOTE In favorable springs the landscape is colored over extensive areas by the showy orange-colored (rarely white or pink) flowers. It is reported that in southern Arizona the plants are grazed by cattle in winter and early spring, when other feed is scarce.

HENRIK KIBAK

AMARANTHACEAE Amaranth Family

DESERT ORACH, CATTLE SALTBUSH
Atriplex polycarpa (Torr.) S. Watson

DESCRIPTION Shrub, 3 feet high, branched throughout, the branches slender. **Leaves** opposite, crowded, small, ¼ to ⅜ inches long, oblong or obovate, entire, dull gray-green and scurfy. Fruit bracts with short linear teeth, these appendaged on the back. **Flowers** greenish, small and inconspicuous.

FLOWERING Spring.

ELEVATION Up to 2,500 feet.

WHERE FOUND Common on sandy-gravelly plains and mesas, tolerant of saline conditions.

NOTE In the deserts of southwestern Arizona it covers vast areas of moderately saline or nonsaline soil, in pure stands or associated with creosotebush, the bushes often symmetrically rounded and evenly spaced. It is by far the most important native forage plant of that region, which, however, is too arid to support cattle in any significant number.

SAM KIESCHNICK

AMARANTHACEAE Amaranth Family

RUSSIAN-THISTLE, TUMBLEWEED
Salsola tragus L.

SYNONYM *Salsola pestifer* A. Nels.

DESCRIPTION Densely branched annual herb 2 to 3 feet high, the branches light-green and somewhat ribbed. **Leaves** awl-shaped, rigid, light-green. **Flowers** green, the calyx five-parted, the segments becoming winged on the back.

FLOWERING Summer to fall.

ELEVATION Low elevations.

WHERE FOUND When the plants die they break off at the ground and become tumbleweeds. This is a very common weed found along roadsides and in cultivated fields.

NOTE Extensively naturalized in the western United States; introduced from Eurasia. In early spring the young plants are readily eaten by livestock, and the dead plants are eaten in winter after softening by rains. The plant is a typical "tumbleweed," breaking off at the surface of the ground when mature and piling up along fences. One of the Hopi Indian names signifies "white man's plant."

RACHEL STRINGHAM

ASTERACEAE Aster Family

BIG BURSAGE

Ambrosia ambrosioides (Delpino) W.W. Payne

SYNONYM *Franseria ambrosioides* Cav.

DESCRIPTION Shrubby, 3 to 4 feet tall, with many coarse stems from the base, these covered with a short rough pubescence. **Leaves** 4 to 6 inches long, lanceolate, irregularly coarsely toothed, dark green. The surface is covered with short rough hairs. **Staminate flowers** in a spike, each flower a round disk ¼ inches across. The numerous stamens with bright yellow anthers are borne on the disk. **Pistillate flowers** composed of small spine tipped bracts, this forms the inchesbur. inches.

FLOWERING Spring.

ELEVATION Up to 3,000 feet.

WHERE FOUND Very common in sandy washes and canyons.

SAM KIESCHNICK

ASTERACEAE Aster Family

SLIMLEAF BURSAGE
Ambrosia confertiflora DC.

DESCRIPTION Annual or biennial, 1 to 2 feet high, branched from the base and somewhat throughout. **Leaves** 1½ to 2 inches long, pinnately divided into lobes which are again divided, somewhat gray-green from the pubescence which covers them. **Inflorescence** spike-like, flowers ⅛ inches long, composed of an involucre filled with several small yellow heads.

FLOWERING Summer.

ELEVATION Up to 3,500 feet.

WHERE FOUND Common on mesas and slopes, sometimes a weed in cultivated land.

NOTE The plant has a strong, heavy odor.

BOB MILLER

ASTERACEAE Aster Family

RABBITBUSH, TRIANGLE-LEAF BURSAGE

Ambrosia deltoidea (Torr.) W.W. Payne

SYNONYM *Franseria deltoidea* Torr.

DESCRIPTION Low rounded shrub, 12 to 24 inches high, branched from the base and throughout, the branches brittle, dark green to dark brown. **Leaves** numerous, silvery-green, 1 to 1½ inches long, ½ to ⅝ inches wide, cordate-acuminate, margins shallowly toothed and often somewhat wavy, surfaces finely rugose, puberulent. **Flowers** numerous, inconspicuous, the staminate composed of disks about ¼ inches across bearing numerous stamens on the margins, the pistillate composed of a cone-like structure of spine tipped scales, when dry these form small burs about ¼ inch long.

FLOWERING Spring.

ELEVATION Up to 3,000 feet.

WHERE FOUND Very common on hillsides, often forming nearly pure stands.

ROBERT DANIEL AVILA

ASTERACEAE Aster Family

BURROWEED, WHITE BURROBUSH

Ambrosia dumosa (A. Gray) W.W. Payne

SYNONYM *Franseria dumosa* A. Gray

DESCRIPTION Rounded shrub 2 to 2½ feet high, branched throughout, the **stems** slender and with a gray pubescence. **Leaves** ½ to 1 inches long, very slender, deeply pinnately lobed and gray-green in color. **Flowers** in terminal spikes 2 to 3 inches long. Staminate and pistillate flowers found on the same spikes. Staminate flowers are small disks covered with tiny yellow stamens, pistillate flowers are surrounded by short, spiny bracts. **Fruits** small yellow-green burs.

FLOWERING Spring.

ELEVATION Up to 3,000 feet.

WHERE FOUND Common on dry plains and mesas.

NOTE This plant is said to be preferred for forage by horses to all other desert shrubs.

INDIO BROWN

ASTERACEAE Aster Family

CHEESEBUSH, WHITE RAGWEED

Ambrosia salsola (Torr. & A. Gray) Strother & B.G. Baldwin

DESCRIPTION Shrub, 1½ to 3 feet high, intricately branched throughout, the branches slender. **Leaves** linear-filiform, light green. **Flower heads** small, numerous, inconspicuous, green or dull yellow-green.

FLOWERING Spring.

ELEVATION Up to 4,000 feet (usually lower).

WHERE FOUND Common in sandy washes and on rocky slopes, sometimes in saline soil.

ANDERS HASTINGS

EPHEDRACEAE Mormon-Tea Family

MORMON-TEA

Ephedra viridis Coville

DESCRIPTION Shrub 3 feet tall and as wide, common but rather inconspic-uous except when in bloom. The **lower branches** are heavy, dull gray and with a soft shredding bark, the **stems** are slender, yellow-green, erect and whip-like, the **leaves** are reduced to two minute colorless pa-pery scales at each joint. The **pistillate flowers** are green and the **sta-minate flowers** yellow.

FLOWERING Spring.

ELEVATION 3,000 to 7,000 feet.

WHERE FOUND Common in creosotebush scrub, rocky slopes.

NOTE A valuable winter browse plant when better forage is lacking. A palat-able tonic beverage (Mormon-tea, Brigham-tea) is made from the dried stems and flowers of these plants, which contain certain alkaloids, such as pseudoephedrin, and tannins (use of which is not recommended today). The Indians and early white settlers esteemed *Ephedra* for treat-ment of syphilis and other affections. The drug ephedrin, commonly administered as an astringent and as a mild substitute for adrenalin, is obtained from *Ephedra sinica*, a Chinese herb.

SULA VANDERPLANK

EUPHORBIACEAE Spurge Family

BEETLE SPURGE
Euphorbia eriantha Benth.

SYNONYM *Poinsettia eriantha* (Benth.) Rose & Standl.

DESCRIPTION Shrubby, 1 to 3 feet high, branched; stem at base ¼ inches thick, with a light tan, somewhat wrinkled, sparsely hairy bark, above yellow-green. **Leaves** 2 inches long, ⅛ inches wide, linear, tapering to a very narrow base, yellow-green, minutely puberulent. **Flowers** inconspicuous in small clusters at the apices of the branches. Flowers yellow-green in color, corolla absent. **Fruit** ³/₁₆ inches long, oblong, dull purple, covered with short white pubescence in an irregular pattern.

FLOWERING Summer to fall.

ELEVATION Up to 4,000 feet.

WHERE FOUND Common on dry hot slopes and canyons.

NOTE The long slender leaves are clustered at the tips of the stems similar to those of the large red Poinsettia.

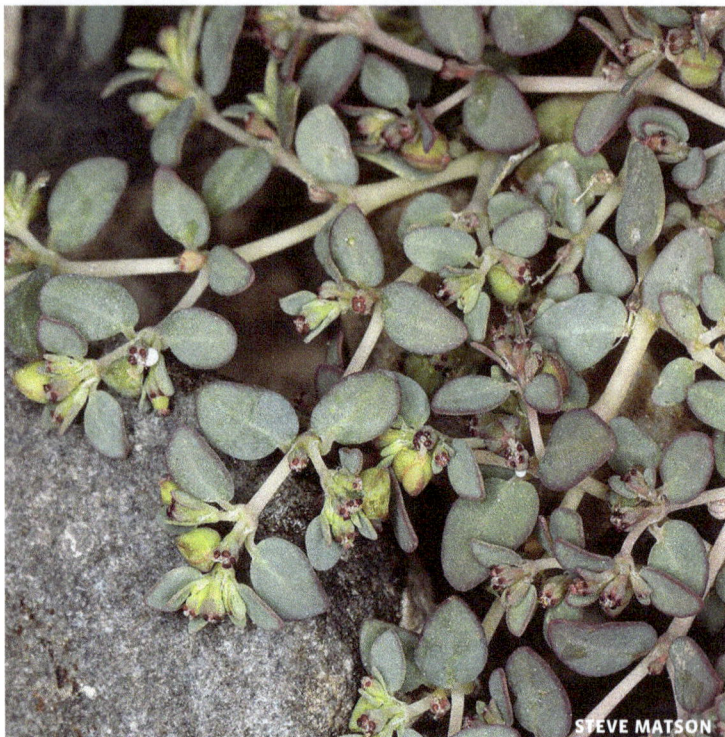

STEVE MATSON

EUPHORBIACEAE Spurge Family

SONORAN SANDMAT
Euphorbia micromera Boiss. ex Engelm.

SYNONYM *Chamaesyce micromera* (Boiss.) Woot. & Standl.

DESCRIPTION Annual herb, 2 to 4 inches high, spreading 4 to 6 inches from the center of the plant; branched throughout, the branches very slender, yellow, breaking easily at the somewhat swollen joints. **Leaves** green to red-brown in color, ⅛ inches long, ovate with an oblique base, petioles $1/_{32}$ inches long, very slender, stipules $1/_{32}$ inches long, white, plumose. **Flowers** small, the four conspicuous petal-like glands about $1/_{16}$ inches long, these brown at the base and white at the margins.

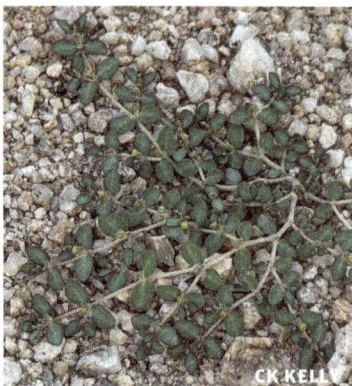

CK KELLY

FLOWERING Spring to summer.

ELEVATION Up to 5,000 feet.

WHERE FOUND Very common on sandy soils.

SAM KIESCHNICK

POACEAE Grass Family

PURPLE THREE-AWN

Aristida purpurea Nutt.

DESCRIPTION Growing 1 to 2 feet high; **head** composed of many purple or lavendar flowers, these are very slender and each bears three long spreading awns. This is an attractive grass when growing in clumps.

FLOWERING Spring to summer.

ELEVATION Up to 5,000 feet.

WHERE FOUND Common on dry rocky hills and plains.

ABRAHAM ROMERO

POACEAE Grass Family

SIX-WEEKS GRAMA
Bouteloua barbata Lag.

DESCRIPTION Annual, tufted with many slender stems from the base. **Spikes** four to six along the main flower stem. The **flowers** are closely crowded on each side of the spike. Flowers about ⅛ inch long, dark purple and each with three awns 1/16 inch long. The spikes are ½ inch long.

FLOWERING Summer.

ELEVATION Up to 5,500 feet.

WHERE FOUND Common on dry mesas and rocky hills.

ROBERT DANIEL AVILA

POACEAE Grass Family

BERMUDA GRASS
Cynodon dactylon (L.) Pers.

DESCRIPTION The **leaves** form a dense mat above which the flower stalk rises 6 to 7 inches. The small **flowers** are crowded along the lower edge of the four or more spreading rays at the summit of the stalk.

FLOWERING Summer.

ELEVATION Low elevations only.

WHERE FOUND Common in lawns and waste places; introduced in America, and the common Bermuda grass of lawns in Phoenix.

PATTY MITCHUM

POACEAE Grass Family

WALL BARLEY

Hordeum murinum L.

DESCRIPTION Stalks mostly solitary, the heads partly enclosed by the upper sheathing leaves; **flowers** closely crowded, each with a long erect awn. Somewhat resembling a head of wheat.

FLOWERING Spring.

WHERE FOUND A common introduced weed in cultivated ground and waste places.

SUE CARNAHAN

POACEAE Grass Family

BIGELOW'S BLUEGRASS

Poa bigelovii Vasey ex Scribn.

DESCRIPTION Plant 12 to 15 inches high, slender. **Leaves** 2 to 2½ inches long, ⅛ inch wide, narrowly linear and with a sharp tip. **Inflorescence** 2 to 4 inches long, the flowers numerous but not densely crowded, ⅛ inch long, yellow-green, the glumes papery, blunt at the apices, and with the margins hairy.

FLOWERING Spring.

ELEVATION Up to 3,000 feet.

WHERE FOUND Common, open ground.

JACK BYRLEY

POLYGONACEAE Buckwheat Family

CANAIGRE DOCK, WILD-RHUBARB
Rumex hymenosepalus Torr.

DESCRIPTION Coarse perennial herb, 2 to 3 feet high. **Leaves** mostly basal, 6 to 12 inches long, broad, curling somewhat, dark green. **Inflorescence** a closely crowded panicle of numerous small green flowers. In fruit the inner sepals or valves become enlarged, somewhat papery and red-brown; these are very conspicuous. The **root** is large and tuberous.

FLOWERING Summer to fall.

ELEVATION Up to 6,000 feet.

WHERE FOUND Common and conspicuous in sandy stream beds, ditches, fields, etc.

NOTE The petioles make a good substitute for rhubarb in pies. Indians and Mexicans used the leaves of this and other species of *Rumex* for greens and eat the petioles roasted or stewed with sugar. As with other greens that are sufficiently succulent, the Papago Indians roasted rather than boil canaigre leaves, probably because of the frequent scarcity of water. The Hopi and Papago Indians used the roots for treating colds and sore throat, and a dye was formerly obtained from them by the Navajos.

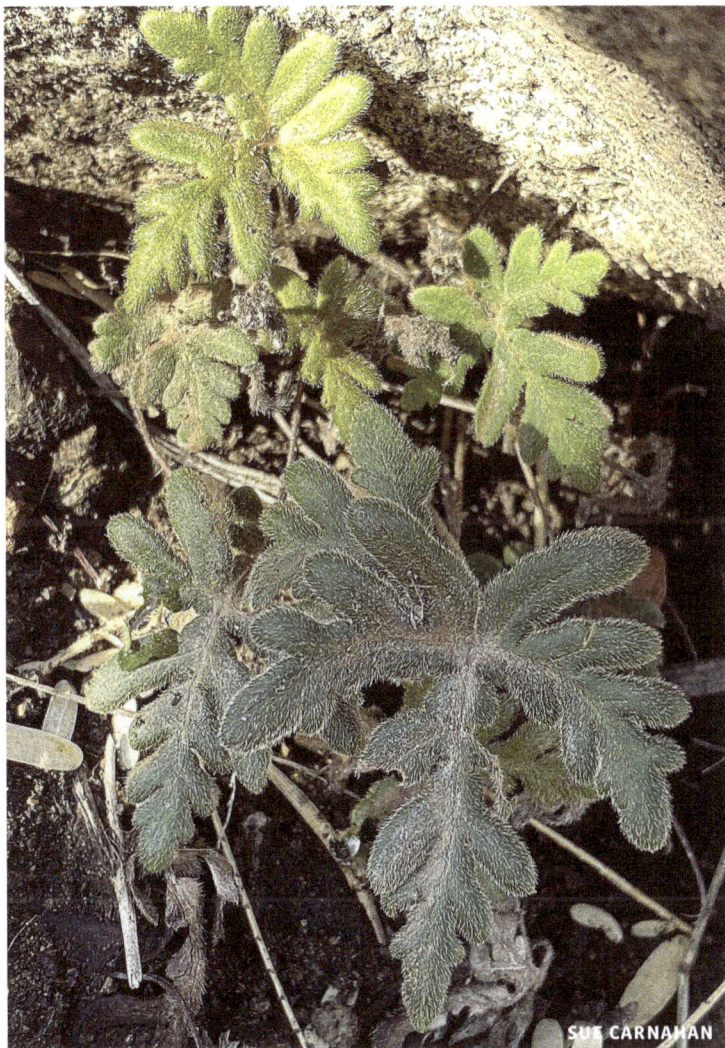

SUE CARNAHAN

PTERIDACEAE Maidenhair Fern Family

COPPER FERN

Bommeria hispida (Mett. ex Kuhn) Underwood

SYNONYMS *Gymnopteris triangularis* (Kaulf.) Underw., *Hemionitis hispida* (Mett.) Christenh.

DESCRIPTION The **fronds** are about 2 inches across, pentagonal in outline, dark green above and with a gold or silver powder covering the under surface depending upon the age of the frond.

ELEVATION 4,000 to 6,000 feet.

WHERE FOUND Moist shaded cliffs, not common.

SUE CARNAHAN

PTERIDACEAE Maidenhair Fern Family

PRINGLE'S LIP FERN
Myriopteris pringlei (Davenport) Grusz & Windham

SYNONYMS *Cheilanthes pringlei* Davenport, *Hemionitis pringlei* (Davenport) Christenh.

DESCRIPTION Growing in damp shady places under rocks. The **fronds** are 3 to 4 inches long, narrow and with the pinnae arranged along the mid-rib. Surfaces light green above, darker and somewhat woolly with short tan hairs below.

ELEVATION 3,000 to 5,000 feet.

WHERE FOUND At base of cliffs, not common.

SUE CARNAHAN

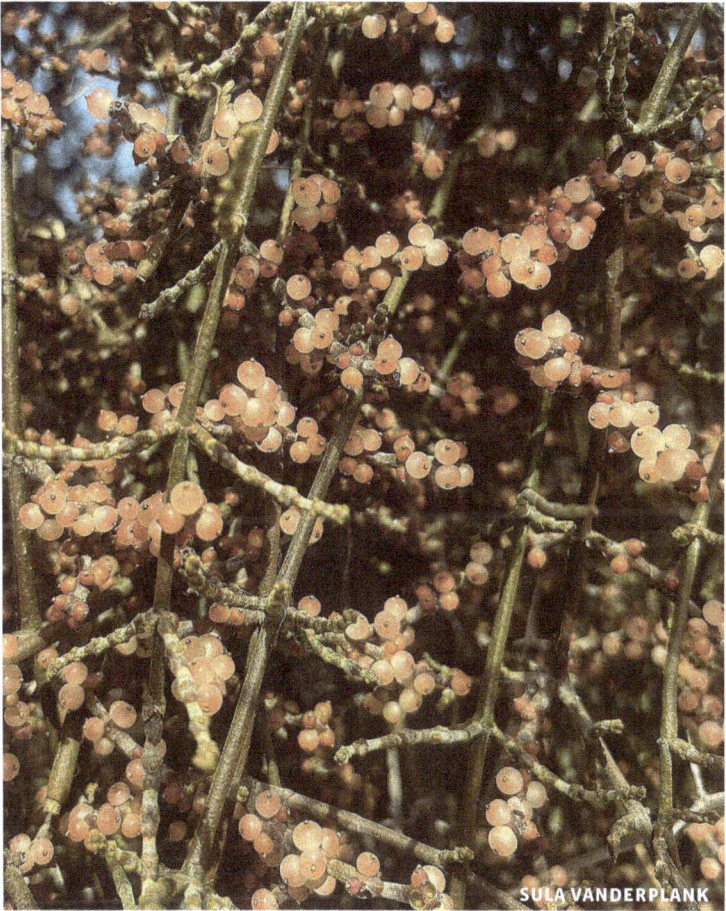

SULA VANDERPLANK

SANTALACEAE Sandalwood Family

MESQUITE MISTLETOE
Phoradendron californicum Nutt.

DESCRIPTION Parasitic on the ironwood and mesquite. **Stems** 8 to 15 inches long or longer, slender, limber, dark yellow-green or red-brown; joints close together, the stems breaking easily at these joints. **Berries** numerous, globose, dull orange-red.

FLOWERING Summer.

ELEVATION Up to 4,000 feet.

WHERE FOUND Parasitic chiefly on leguminous shrubs and trees (*Acacia, Prosopis, Cercidium*), but occasionally on members of Buckthorn Family (Rhamnaceae); common.

RACHEL STRINGHAM

SELAGINELLACEAE Selaginella Family

ARIZONA SPIKEMOSS

Selaginella arizonica Maxon

DESCRIPTION Low-growing moss-like plants, 2 to 3 inches tall, spreading, the stems rooting on the under side. **Leaves** scale-like, dark green and glossy, about ⅛ inch long. **Spores** borne on small erect cones. These cones are similar to the vegetative branches in color but are four-sided, the minute orange spores being borne in the axils of the leaves.

WHERE FOUND Found among the rocks on the slopes in the canyons; common.

NOTE During dry seasons, plants appear to be dead but will revive quickly following rain or when put in water.

GLOSSARY

Achene. A small, dry, hard, 1-seeded fruit.

Alternate (leaves). Borne singly at a node, first on one side of the stem, then on the other.

Annual. Living a single season.

Anther. The pollen-producing part of the stamen.

Apex. Upper end or tip (of leaf, petal, etc.).

Appressed. Lying flat and close against.

Aril. A fleshy, often bright-colored, appendage to a seed.

Auricle. Ear-shaped appendage or lobe.

Awn. Slender, usually stiff, terminal bristle.

Axil. Angle formed between the leaf and stem, or branch and stem.

Axillary. In an axil.

Beak. A projection ending in an elongated tip.

Bearded. Bearing long or stiff hairs.

Berry. A fleshy fruit having a thin skin or outer covering, the seeds surrounded by the pulp.

Biennial. Living only two years and flowering the second year.

Bisexual. Having both stamens (male organs) and pistils (female organs).

Bract. A small leaf, often scalelike and usually subtending a flower or inflorescence.

Bracted. Having bracts.

Bulb. Underground leaf-bud with fleshy scales.

Calyx. The outer, usually green, portion of the flower.

Capitate. Headlike; arranged in a head, or dense cluster.

Capsule. A dry fruit composed of several cells, opening at maturity.

Carpel. A simple pistil, or one member of a compound pistil.

Chaff. The scales or bracts on the receptacle of many composite flowers.

Chlorophyll. The green coloring matter of plants.

Ciliate. Fringed with hairs.

Clasping. Partly surrounding another structure at the base, e.g., as a leaf surrounds a stem.

Clavate. Club-shaped; gradually thickened upward.

Compound. Composed of 2 or more similar parts united into one whole. **Compound leaf**, one divided into separate leaflets.

Connective. The part of a stamen that joins the two anther cavities.

Cordate. Heart-shaped, the broadest part at the base.

Corm. The enlarged, solid, bulblike base of some stems.

Corolla. The second or inner set of floral parts, composed of separate or united united petals; usually conspicuous by its size and/or color.

Corona. A crownlike structure on inner side of the corolla, as in narcissus, milkweed, etc.

Corymb. A flat-topped or convex flower cluster, the outer flowers opening first.

Crenate. Having much-rounded teeth.

Cuneate. Shaped like a wedge.

Cyathium. Cuplike involucre around the flowers in *Euphorbia*.

Cylindric. Shaped like a cylinder.

Cyme. A convex or flat flower cluster, the central flowers unfolding first.

Decumbent. Stems in a reclining position, but with the end ascending.

Decurrent. Extending downward (applied to leaves in which the blade is extended as wings along the petiole).

Dentate. With sharp teeth.

Diadelphous (referring to stamens). Combined in two, often unequal, sets.

Dicotyledon or **Dicot**. Plant with 2 seed-leaves or cotyledons. In practice usually recognized by the net-veined leaves, the parts of the flower in 4's or 5's, the fibers of the stem in a ring.

Disk. Central part of the head of many composite flowers.

Disk-flower. In composite flowers (Asteraceae), the tubular flowers in the center of the head, or comprising the head.

Dissected. Cut or divided into narrow segments.

Drupe. A fleshy fruit with a single hard stone in the center, such as the peach, plum, or cherry.

Ellipsoid, Elliptic. Oval in outline, widest at the middle and narrowed about equally at the ends.

Entire. With a continuous, unbroken margin; without teeth, scallops, or lobes.

Fertile. Capable of normal reproductive functions. A fertile anther produces pollen; a fertile flower produces fruit.

Filament. The stalk of the stamen.

Filamentous. Threadlike.

Flexuous. Curved alternately in opposite directions.

Foliate. Having leaves, leaved, as 3-leaved.

Foliolate. Having leaflets.

Follicle. A dry fruit developed from a single ovary, opening by a slit along one side.

Fruit. The seed-bearing product of a plant; the ripened ovary with such other parts as may be attached to it.

Glabrous. Smooth, in the sense of lacking hairs (not necessarily smooth to the touch).

Glaucous. Covered with a fine bluish or whitish bloom.

Globose, globular. Spherical or nearly so.

Glutinous. Gluelike, sticky.

Hastate. Shaped like an arrowhead, but with the basal lobes pointed outward.

Herbaceous. Soft, not woody; leaflike in color or texture.

Hip. The fleshy, ripened fruit of the rose.

Hood. A cap-shaped or hood-shaped structure formed by the sepals and petals in orchids.

Horn. An incurved pointed projection.

Hypanthium. A cuplike receptacle that bears the flower parts on its upper margin.

Inferior. Lower or below (an inferior ovary is one that is seemingly below the calyx segments).

Inflorescence. A flower cluster.

Involucral. Belonging to an involucre.

Involucre. A whorl of small leaves or bracts below a flower or flower cluster.

Irregular (flower). Showing inequality in the size, form, or union of its similar parts.

Keel. The two fused lower petals of the flower of the Pea Family.

Keeled. Having a central ridge.

Lanceolate. Shaped like a lance-head (much longer than wide, widest near the base and tapering to the apex).

Leaflet. One of the divisions of a compound leaf.

Legume. A fruit from a simple ovary, opening along 2 sides; e.g., a pea pod.

Ligulate. Furnished with a ligule (applied to the ray flowers in the composites).

Ligule. A straplike organ.

Limb. The expanded part of a corolla of united petals.

Lip. The upper or lower division of an irregular corolla or calyx, as in the mints. The characteristic (and apparently lower) petal of an orchid is called the lip.

Lobe. A partial division of a leaf or other structure.

Lyrate. Pinnately cut, with a large, rounded lobe at apex, but with the lower lobes small.

Membranous. Thin and somewhat transparent, like a membrane.

Midrib. The central rib or vein of a leaf.

Monocotyledon or **Monocot**. A plant with 1 seed-leaf (cotyledon). In practice usually recognized by the parallel veins of the leaves, the flower parts in 3's or 6's, the fibers of the stem scattered (not in a ring).

Naked receptacle. Without scales or bracts.

Nectar. A sweet, often fragrant liquid.

Nectariferous, or **Nectiferous**. Producing or having nectar.

Nectary. Any place or organ where nectar is secreted.

Nerve. A principal, unbranched vein.

Net-veined. Having veins forming a network or reticulum.

Node. A point on a stem where leaves, branches, or buds are attached.

Nutlet. A small hard, 1-seeded fruit.

Ob-. Prefix meaning inverted, e.g., obovate, meaning ovate but with the broadest part toward the apex.

Oblong. Having the length two to three times the width, the sides nearly parallel.

Obtuse. Blunt or rounded at the end.

Opposite (leaves). A pair of leaves arising from the same node and on opposite sides of the stem; of stamens, inserted in front of the petals, and thus opposite or across from them.

Ovary. The part of the pistil that produces seeds.

Ovate. Egg-shaped, the broadest part basal.

Palmate. Having parts diverging from a common base, as the fingers from a hand.

Panicle. A loose, elongate, flower cluster with compound branching.

Papilionaceous. Having a standard, wings, and a keel and resembling a butterfly, as in a pea flower.

Pappus. The calyx in the Asteraceae, consisting of hairs, bristles, awns, or teeth at the top of the ovary or achene.

Parallel (venation), **parallel-veined**; having veins rising at base of a leaf and continuing to apex in a nearly parallel manner.

Parasite. A plant that gets its nourishment from another living plant to which it is attached.

Parted. Cut or lobed more than halfway to the middle or base.

Pedicel. The stem of a single flower in a flower cluster.

Peduncle. The stem or stalk of a single flower, or of a cluster of flowers.

Perennial. A plant that is able to live year after year.

Perfect (flower). Having both stamens and pistils.

Perfoliate. Having the stem apparently passing through the leaf.

Perianth. The flower envelope, consisting of the calyx and the corolla (if present).

Petal. One of the parts of the corolla.

Petiole. The stem or stalk of a leaf.

Pinnate. Like a feather, having the parts arranged in two rows along a common axis.

Pistil. The seed-bearing part of the plant, produced in the center of the flower.

Pistillate. Provided with pistils.

Plumose. Featherlike, with fine, soft hairs along the sides.

Pod. A simple, dry fruit which opens when ripe; a legume.

Pollen. The spores (male) produced by the anthers.

Prostrate. Lying flat on the ground.

Pubescent. Hairy, the hairs soft.

Punctate. Covered with dots or pits.

Raceme. A flower cluster with pedicelled flowers borne along a somewhat elongated common axis, the lower flowers opening first.

Ray. One of the branches of an umbel; a strap-shaped flower in a head in which tubular disk flowers are also present.

Receptacle. The end of a flower stalk upon which the flower parts are borne, or the enlarged end of the peduncle upon which the flowers are borne in the Aster Family.

Recurved. Curved downward or backward.

Reflexed. Bent downward or backward.

Regular (flower). Having all members of each kind similar in size and shape, e.g., the petals all alike.

Reticulate. Like a network.

Rhombic. Somewhat diamond-shaped.

Rib. A primary or prominent vein of a leaf.

Rootstock. (also called **rhizome**). A prostrate or underground stem, usually rooting at the nodes.

Rosette. A cluster of leaves in circular form, usually at the base of a plant.

Rotate (corolla). Wheel-shaped, having a short tube and widely spreading limb, as in the forget-me-not.

Runner. A horizontal, above-ground stem that may root and develop new plants.

Sagittate. Shaped like an arrowhead.

Salverform. Having a slender tube (relatively long) and widely spreading limb.

Scape. A leafless, flowering stalk arising from the ground or from a very short stem bearing basal leaves.

Seed. A mature ovule which can germinate and give rise to a new plant.

Sepal. One of the parts of the calyx or outer floral envelope.

Sessile. Lacking a stalk or stem.

Sheath. A tubelike part which surrounds another part, as a leaf may sheathe a stem.

Shrub. A plant, shorter than a tree, with several to many woody stems which do not usually die at the end of the growing season.

Simple. Not compound; composed of a single piece or unit.

Sinus. The indentation between two lobes.

Solitary. Single.

Sordid. Dirty white, dingy.

Spadix. A spike of flowers on a fleshy axis.

Spathe. A large, often showy bract enclosing or subtending a flower cluster (usually a spadix).

Spatulate. Shaped like a spatula, oblong with a long, somewhat tapering base, the apex rounded.

Spike. An elongated flower cluster with sessile or nearly sessile flowers borne along a common axis.

Spur. A hollow, somewhat pointed projection.

Stamen. The part of the flower that bears the pollen.

Staminate. Having stamens.

Staminode. A sterile stamen; a stamen that does not bear pollen.

Standard. The upper, enlarged petal of a pea flower.

Stem. The major supporting system of a plant to which buds, leaves, and flowers are attached.

Sterile. Unproductive, as a flower without a pistil, or a stamen without an anther.

Stigma. The part of a pistil upon which pollen germinates.

Stolon. A basal branch rooting at the nodes to produce new plants; a runner.

Stoloniferous. Bearing or producing stolons.

Style. The portion of the pistil between the ovary and stigma.

Sub-. Prefix meaning somewhat or slightly, e.g., subcordate, meaning somewhat heart-shaped.

Subtend. To extend under, as a bract or involucre subtends a flower or flower cluster.

Succulent. Juicy or fleshy in texture, usually resistant to drying.

Superior (ovary). Placed above the point of attachment of the other flower parts.

Synonym. A former scientific name of a plant.

Tendril. A slender, almost thread-like, twining structure.

Throat. The part in a corolla or calyx where the tube and limb come together.

Tooth, teeth. Small, sharp-pointed projection or projections.

Tuber. A thickened, fleshy, modified, underground stem having numerous buds or eyes, e.g., the Irish potato.

Tubercle. A small, swollen, and usually hardened structure.

Tubular. Shaped like a tube.

Tufted. Forming clumps.

Twining. Winding spirally.

Umbel. A flower cluster with all the pedicels arising from the same point.

Undulate. Having a wavy margin or surface.

Unisexual. Of one sex, either staminate or pistillate only.

Urn-shaped (corolla). Having an enlarged, globular base, narrowed at the neck, and with a small limb.

Vein. A thread of visible fibrovascular (transporting) tissue.

Venation. Character or pattern of the veining.

Whorl. A group of three or more similar parts in a circle around a stem.

Winged. A thin expansion of tissue as along a leaf petiole. Also refers to a lateral petal as in a pea flower.

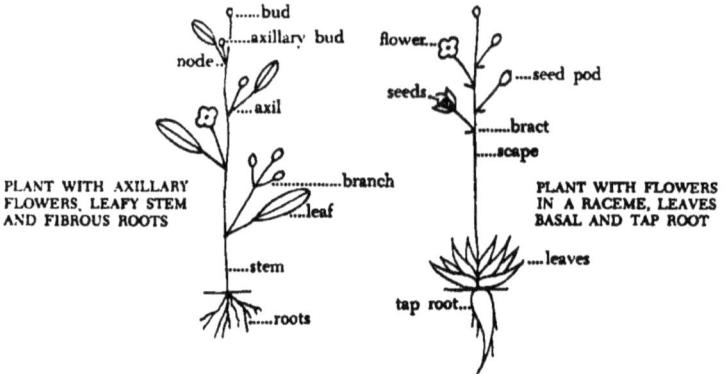

PLANT WITH AXILLARY
FLOWERS, LEAFY STEM
AND FIBROUS ROOTS

bud
axillary bud
node
axil
branch
leaf
stem
roots

flower
seed pod
seeds
bract
scape

PLANT WITH FLOWERS
IN A RACEME, LEAVES
BASAL AND TAP ROOT

leaves
tap root

PLANTS AND THEIR PARTS

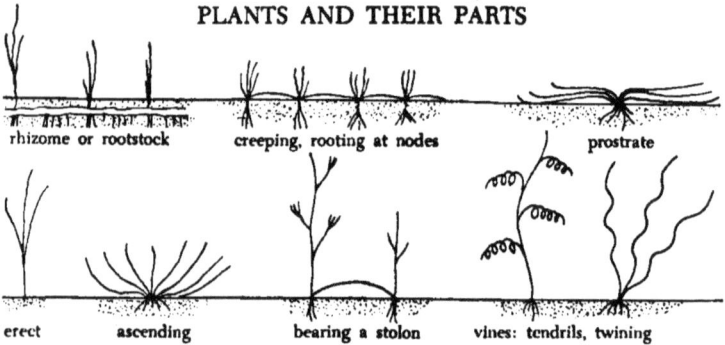

rhizome or rootstock   creeping, rooting at nodes   prostrate

erect   ascending   bearing a stolon   vines: tendrils, twining

GROWTH PATTERNS OF STEMS

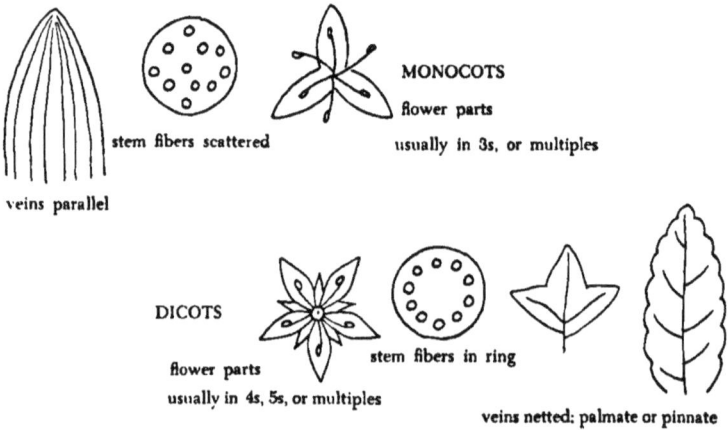

veins parallel

stem fibers scattered

MONOCOTS

flower parts

usually in 3s, or multiples

DICOTS

flower parts

usually in 4s, 5s, or multiples

stem fibers in ring

veins netted: palmate or pinnate

MONOCOT AND DICOT CONTRASTS

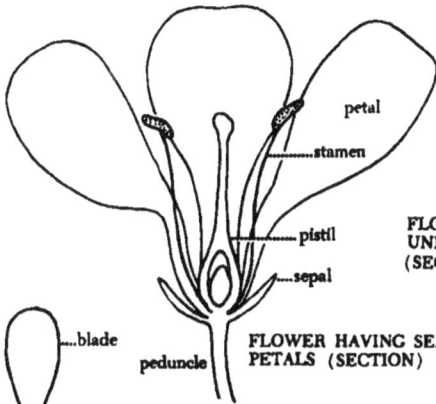

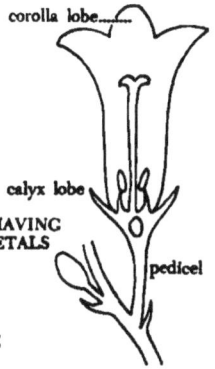

**FLOWER HAVING
UNITED PETALS
(SECTION)**

**FLOWER HAVING SEPARATE
PETALS (SECTION)**

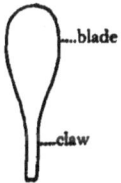

PETAL

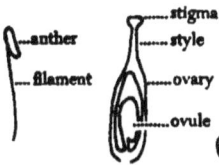

STAMEN   PISTIL

OVARY SUPERIOR

OVARY INFERIOR

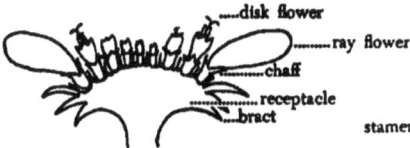

**HEAD OF COMPOSITE
INFLORESCENCE
(SECTION)**

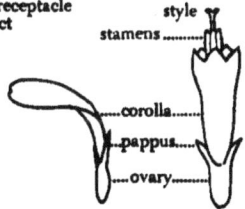

**RAY
FLOWER**   **DISK
FLOWER**

OF A COMPOSITE

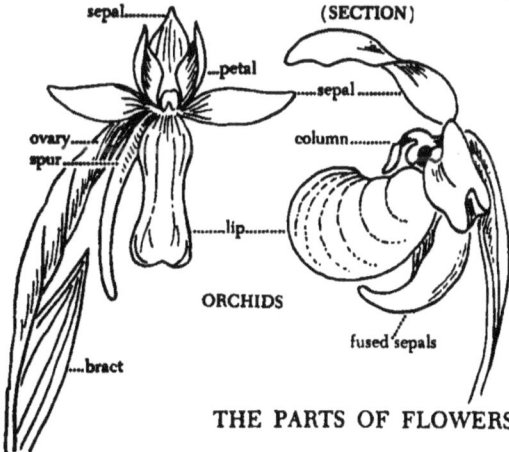

ORCHIDS

THE PARTS OF FLOWERS

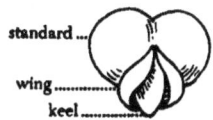

PEA FLOWER

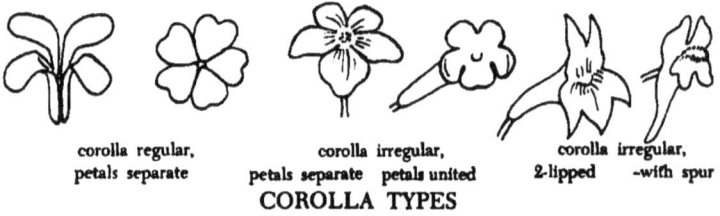

corolla regular,
petals separate

corolla irregular,
petals separate   petals united

corolla irregular,
2-lipped   -with spur

COROLLA TYPES

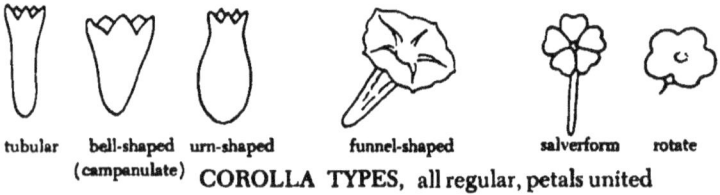

tubular   bell-shaped   urn-shaped   funnel-shaped   salverform   rotate
          (campanulate)

COROLLA TYPES, all regular, petals united

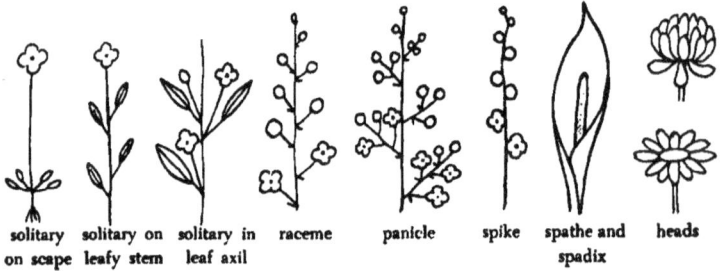

solitary   solitary on   solitary in   raceme   panicle   spike   spathe and   heads
on scape   leafy stem   leaf axil   spadix

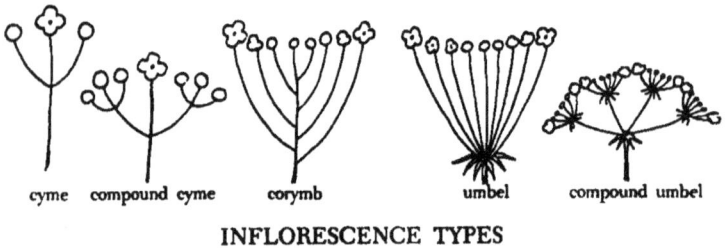

cyme   compound cyme   corymb   umbel   compound umbel

INFLORESCENCE TYPES

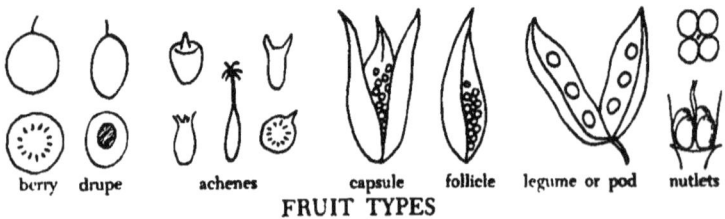

berry   drupe   achenes   capsule   follicle   legume or pod   nutlets

FRUIT TYPES

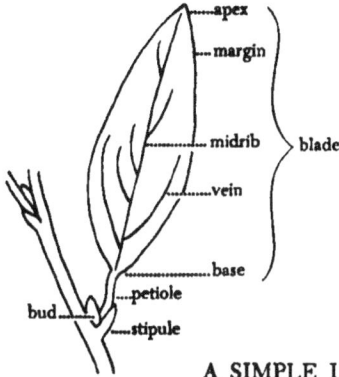

A SIMPLE LEAF

apex
margin
midrib
vein
base
petiole
stipule
bud
blade

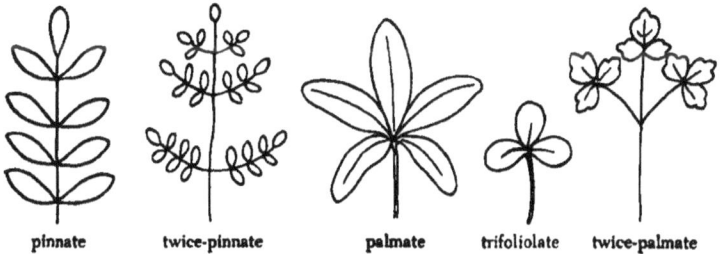

pinnate   twice-pinnate   palmate   trifoliolate   twice-palmate

COMPOUND LEAVES

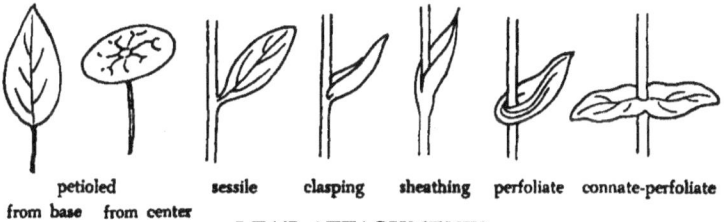

petioled
from base   from center   sessile   clasping   sheathing   perfoliate   connate-perfoliate

LEAF ATTACHMENTS

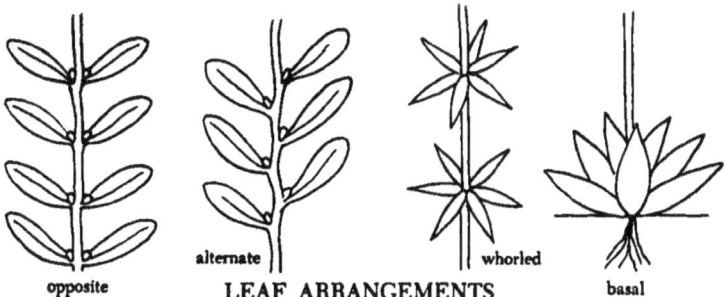

opposite   alternate   whorled   basal

LEAF ARRANGEMENTS

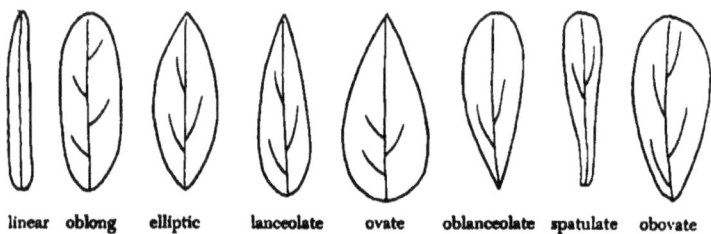

linear oblong elliptic lanceolate ovate oblanceolate spatulate obovate

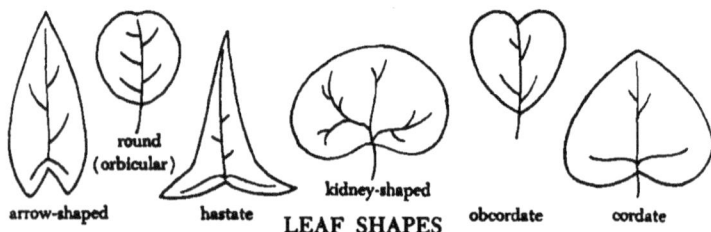

round (orbicular)

arrow-shaped hastate kidney-shaped obcordate cordate

LEAF SHAPES

pointed obtuse rounded truncate obcordate mucronate

LEAF APICES

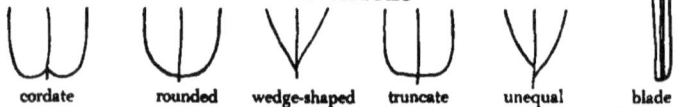

cordate rounded wedge-shaped truncate unequal blade decurrent on petiole

LEAF BASES

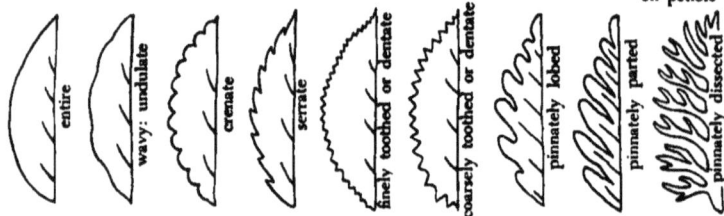

entire wavy: undulate crenate serrate finely toothed or dentate coarsely toothed or dentate pinnately lobed pinnately parted pinnately dissected

LEAF MARGINS

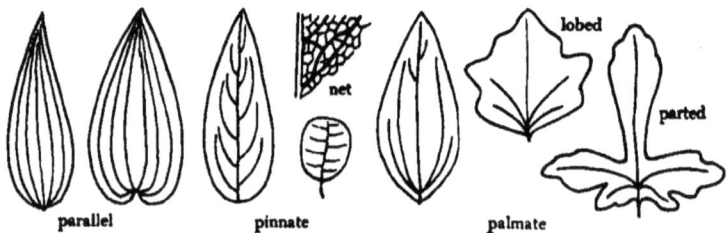

net lobed parted

parallel pinnate palmate

LEAF VENATION

INDEX TO COMMON & SCIENTIFIC NAMES

NOTE: Synonyms are listed in *italics*.

www.ingramcontent.com/pod-product-compliance
Lightning Source LLC
Chambersburg PA
CBHW052113030426
42335CB00025B/2964